D0331227

9/20

What Do Animals Think and Feel?

What Do Animals Think and Feel?

An Investigation into Emotional Behavior

Karsten Brensing

PEGASUS BOOKS
NEW YORK LONDON

WITHDRAWN

FREMONT PUBLIC LIBRARY DISTRICT
1170 N. Midlothian Road
Mundelein, IL 60060

WHAT DO AMIMALS THINK AND FEEL?

Pegasus Books, Ltd.
148 West 37th Street, 13th Floor
New York, NY 10018

Copyright © 2020 by Karsten Brensing

First Pegasus Books hardcover edition October 2020

All rights reserved. No part of this book may be reproduced
in whole or in part without written permission from the publisher,
except by reviewers who may quote brief excerpts in connection with a review
in a newspaper, magazine, or electronic publication; nor may any part of this
book be reproduced, stored in a retrieval system, or transmitted in any form or
by any means electronic, mechanical, photocopying, recording, or
other, without written permission from the publisher.

ISBN: 978-1-64313-554-0

10 9 8 7 6 5 4 3 2 1

Printed in the United States of America
Distributed by Simon & Schuster
www.pegasusbooks.com

FREMONT PUBLIC LIBRARY DISTRICT

As a child, in my dealings with animals I was always guided by the golden rule my parents taught me:

'*Was du nicht willst, das man dir tu, das füg auch keinem andern zu.*'

(Do as you would be done by.)

Contents

I

What bowls me over (or, put more simply, Introduction)

DOLPHINS CALL one another by name and orcas inhabit a culture that is over 700,000 years old. Chimpanzees wage strategic conflicts, whereas bonobos delight in dirty talk. Humpback whales follow the dictates of fashion, while fish use tools and play with thermometers. Rats are avid party animals and crows like to toboggan on snow-covered roofs. Ants recognize themselves in mirrors and spruce themselves up before they return home. Ducklings can pass complicated tests in abstract thinking and snails voluntarily take a spin on a hamster exercise wheel. Dogs punish disloyalty, though they are also capable of forgiveness if you apologize to them. Spiders choose their occupation on the basis of their personality and their individual preferences. And humans, meanwhile, stand and scratch their heads in bafflement at all this.

What on earth is going on with these animals? Hardly a week goes by without some report or other appearing in the press about animals' astonishing aptitudes. We are amazed and surprised, but what makes animals tick and conditions their thought processes remains a closed book to us.

My best friend ever to date was a dog. Oh dear, how sad! I hear you thinking to yourselves, and I grant you, it does indeed sound a bit pathetic; what does that say about my social life? Happily, though, that's not the subject of this book – which is about animals and in large part also about *their* incredible social lives, with colleagues, friends, relatives, enemies and strategically planned territorial wars. The enigma that should engage our interest here is played out exclusively in animals themselves, or to be more precise within their nervous systems, and hence it cannot be directly perceived by us humans. Of course, it is easy enough for us to observe animal behaviour and to draw conclusions from what we have seen, but we cannot simply quiz them on whether we've got it right.

Instead, it may even be the case that our ideas rarely have anything to do with the real situation, as the following example – which was humiliating for me – demonstrates: I had grown up with *Flipper*, and it was a childhood dream of mine to one day have the opportunity to swim with dolphins. And so I chose to write my doctoral thesis in ethology on the interaction between humans and dolphins in swimming lessons and so-called dolphin therapy. In my pilot study – something all canny research students conduct to begin with in order to make sure they're not barking up completely the wrong tree – relying on what I had seen with my own eyes, I wrote that dolphins in swimming teaching programmes clearly seek out the company of humans in the water. Indeed, the endless numbers of people offering swimming therapy use this very claim to promote their business. After a year of video observation and detailed evaluation of the results, however, precisely the opposite turned out to be the case: the dolphins were actually trying – with a demonstrably high level of statistical significance – to avoid the swimmers, no easy feat in a pool that was only as large as the designated children's swimming area in municipal indoor swimming baths. This not only punctured a naive childhood dream but also placed a question mark over my planned career at the time. How could I have got it so badly wrong?

Some decades ago, a pod of orcas was trapped for a dolphinarium in British Columbia. The three whales were fed with fish like all the other animals at the facility, but refused everything they were offered. The management were faced with a choice: should they wait and see what transpired or return the animals to the wild? They waited – until one of the orcas starved to death, which prompted the others to start eating the fish. Had the whales refused out of protest at the people who had abducted them, or was it just that they didn't like fish? In the same region, the spillage from the grounding of the oil tanker *Exxon Valdez* in 1989 brought several orca populations close to extinction, but they did not mate with passing orca pods. This behaviour flew in the face of all the logic of evolutionary theory, according to which the animals should have been overjoyed and fallen upon the visiting whales in order to increase their gene pool. Nowadays we know more: these animals live within a culture that is more than 700,000 years old, with a code of behaviour that prevents one individual from having anything to do with the mammal-consuming killer whales within their species, and stops another from eating fish. The global financial crisis and the behaviour of stock-market players shook the world, and the misconduct of a handful of individuals sent whole economies spiralling into a crash. But surprisingly, the real cause of the crisis did not lie in the greed of a few, as we were given to understand, but in irrational patterns of behaviour that go back over 30 million years, which we share with other primates.

So, there are insects that use tools, fish that inhabit a culture, dolphins that give themselves names, elephants that bury their dead, animals that play fair or deliberately lie, and animals that treat themselves with antibiotics or that get us humans to work for them. But what does it mean when a raven can mentally put itself in the position of another raven in order to predict how it will react, or when a magpie recognizes itself in a mirror, or when mammals in general manage to score just as well in cognitive research tests as human beings? How are we to classify this ability – where, for instance, do animals stand as compared to humans, and when and under what circumstances have we all acquired these aptitudes?

I will endeavour to answer these and similar questions that arise by citing countless studies. When I have finished, you will no doubt ask: so what distinguishes us from animals? Not a lot, as it turns out, but this much I can reveal straight away: we humans have one tiny peculiarity that underpins our success as a species, and that is our ability to use language. As such, then, nothing is about to topple humanity from its throne but even so, in the world you will inhabit after reading this small volume, you and other human beings will no longer be alone; instead, you will be living alongside other self-aware and sentient beings, and who knows, perhaps hereafter you'll exchange a polite greeting with this or that raven you chance upon in your neighbourhood.

II

Going at it like animals

A WHILE AGO I became interested in everything to do with expanded consciousness, and in my search for the philosopher's stone, I stumbled upon tantrism. You know, those Indian sex practices involving odd positions and enlightenment through delayed ejaculation. So at some point I found myself in Dussmann's bookshop in Berlin, boldly plucking a book on the subject from a well-organized shelf of volumes. And while I was genuinely interested in the consciousness-raising aspects of tantrism, my eye was inevitably drawn first to the topic of sex in the Table of Contents. Surprisingly though, I only struck lucky on page 83 of Chapter 5. Now, take a guess what the author had written there? 'Yeah, I started reading the book at this page, too.' I had been rumbled...

Although the writer had tried really hard to present tantrism in the correct light and play down the Western sex-obsessed viewpoint, the topic of sex held me in its thrall: sex sells! Be it on the cover of a TV listings magazine or draped across the bonnet of an SUV, bare flesh invites us to ogle, and more. Sex is an incredibly powerful source of inner motivation, and what's more it's fun. Just as well, really, otherwise our ancient forebears would have done nothing but laze around in the trees and spend their time picking their noses and staring into space, meaning we would never have existed.

But before we launch into a discussion of dirty talk, primates using animal sex toys, the incredible sexual practices that are played out in suburban front gardens, and the intellectual *tours de force* performed by dolphins in order to gain carnal satisfaction, we need to clear up a few basics.

Sex is one of the oldest and possibly the most important of all Mother Nature's inventions, even older than the invention of genders, and has only one drawback: the life it brings into being inevitably ends in death.

But first things first. Simple, single-celled organisms reproduce by means of division. During the process of so-called mitosis, an identical copy of the genetic material is produced and then passed on to the daughter cell, which is to all intents and purposes a clone. Ultimately what this means is a complete, unbroken chain right back to the origins of life. In this instance, then, one can justifiably go so far as to talk in terms of immortality. Maybe next time you're enjoying a glass of cloudy wheat beer or Merlot, you'll spare a respectful thought for the industrious yeast cells, millions of years old, that have given your drink its proper oomph.

Multicellular organisms learned very early on that it was a good idea to exchange genetic material with other 'individuals'. And so it was that even the most primitive multicellular organisms, which were already around when there was still no distinction between animals and plants, doubled their genetic information and sent it on its way. These one-celled gametes had the clear task of uniting intimately with other one-celled gametes in order to intermix their genetic material. And so sex was invented. Yet this intermingling of genetic material also meant that the chain of immortality was broken, since every new organism was a little bit different from both its parents. Sex can therefore be seen as the first deadly sin.

Nonetheless, this strategy had enormous benefits, since the combination of all these various bits of genetic material gave rise to a spate of mutations – that is, to alterations of the original genetic material. The key for life on Earth to flourish in all its glory and biodiversity had been found. To the credit of bacteria, though, we should also add at this point that they too have something similar to sexual reproduction, a process known as conjugation. They exchange small snippets of their genetic material with so-called conjugative pili (or sex pili). If the encoded information makes any kind of sense, then this data, which contains a specific aptitude, will be passed on during division to the daughter clone. If not, it may not cause any disruption, but that will be the end of the bacterium. No risk, no fun.

Yet in the first few million years of life on Earth, sex was not yet so fully developed. It was perfectly possible for the sperm cells of a stem organism to cross with themselves and thereby frustrate the entire effort. Indeed, in principle everything still turns on this problem even nowadays, and the spectrum of measures taken ranges from invention of the sexes via the modest covering of seeds among plants to intricate social rules that lay down who may or may not mate with whom. The last of these strategies is not merely a part of our social culture but is also a driving force behind the genesis of the most complex social networks observable in the animal kingdom. Thus, even up until very recently, the formation of third-order alliances was regarded as a purely human domain – before we gained a better understanding of the behaviour of dolphins off the west coast of Australia, of which more later.

1. Alien sex

Let us begin with something closer to home: namely the sex life of our immediate neighbours and tenants.

I am looking directly out of the window of my study at my garden, and I'm willing to bet you have no idea what bizarre sexual practices perhaps the most incredible animal on our planet is indulging in right before our very eyes and most likely in your herbaceous borders or window box. I'm talking about the water bear or tardigrade, a cute little creature just half a millimetre long that is found pretty much everywhere on Earth, in water and also in low Earth orbit. Like the chordates, a group that includes not just us humans but also fish, birds and reptiles, tardigrades even comprise their own phylum, with around 1,000 different species.

If you had eyes like a microscope, you would from time to time see the following scenario played out in your garden: a lover stealing up clandestinely on a female with the aim of seducing her. After a spell of paying court to her, he finally moves in to clinch an intimate kiss. This kiss is not only the prototype of a French kiss; no, it's much more than that, since this kiss is also the actual moment of mating. Scientists believe that the male sperm is transmitted in this act. It is thought that the female carries this sperm with her until the next time she sheds her skin, laying both it and her eggs together in her former exoskeleton, which now forms a protective covering for her brood. Alternatively, it may simply be that she swallows the sperm. There are even some genuinely hermaphroditic species that are both male and female. Unfortunately, nobody knows whether – and, if so, how – these species kiss themselves. Alongside these instances of tenderness, though, there are also examples of brutal violation. In this form of mating, the groom does not tear off the bride's clothes but instead rips open her belly and simply stuffs his semen inside without more ado.[1] It's just another method of internal fertilization. Now, I imagine you're thinking that, for sure, nothing of the sort would ever happen in your bed. But you'd be wrong; the common bedbug, a classic human parasite, goes about things in much the same way. The male bedbug's sharp phallus, which is shaped like a hypodermic needle, punctures the female's abdomen and pumps the sperm into her bloodstream.[2]

Given all these different approaches, it is perhaps unsurprising that some species of tardigrades are also capable of asexual reproduction, so-called parthenogenesis, and have foresworn sex in any shape or form. In this process, the females simply inseminate themselves, without being hermaphroditic. In order for this to function, the animal's body is duped by its own hormones into thinking that it is carrying a fertilized egg. This same trick is also shared by certain species of reptile and worms. Of course, it's evolutionarily problematic, and no doubt dispiriting for males, but if I am a female that has ended up in a remote place without males, it is a great advantage to me if I can set to work building up a new population all on my own. When this population has grown large enough, often a number of males will be produced, and the fun can begin all over again.

I'm conscious of the fact that I still need to explain the mystery of how a proper multicellular animal with a nervous system, muscles and digestive system can get into low Earth orbit and survive there. Water bears are true survival artists. You can heat them to over 250°C (482°F), subject them to the extreme pressures of the deep ocean, and even shoot them into space, and still not kill them. The last of these factors was demonstrated by the biologist Bob Goldstein from the University of North Carolina, when he sent tardigrades into space aboard a Soyuz rocket for ten days, exposing them to both a complete vacuum and solar radiation. Back on Earth, he managed to revive most of these small astronauts in their natural spacesuits using nothing but a drop of water. Their trick was this: as the funny little creatures slowly dry out, they surround themselves with a kind of capsule that protects them and inside which they are able for all practical purposes to shut down all their vital functions. Small wonder that some researchers are already speculating whether tardigrades might have arrived on Earth with meteorites.

But other species also have some truly explosive surprises in store regarding their sex lives. You will no doubt all be familiar with that climactic moment when those involved in sexual congress emit a blissful moan. Happily for us, it is not our dying breath as it is for the unfortunate male bee. It is a well-known fact that male bees, which are called drones, die after the sex act. But do you know how they die? Well, they explode, and their spectacular suicide even has a signalling effect. Drones are utterly captivated by the phero-mones emitted by the queen, a scent that is designed to attract them. Once they pick it up, they turn themselves into the wind and literally follow their nose to the alluringly scent-emitting and receptive queen bee. They home in on her while still in flight and their reproductive organs latch fast on to her body. This moment marks the end of the male bee; he stiffens and the queen, through the muscular contraction of her lower body, draws him closer to her. She does this with such fervour that the poor drone often bursts with an audible pop.[3] All that remains of him thereafter is his sex organ, stuck in the queen's lower body. Certain species, like the mole and a few rodents, leave behind a kind of chastity belt after the sex act by sealing the female's copulatory opening with a sticky bung. Not so our drones: the male's endophallus, which appears to us as orange-coloured tissue, but which also glows in the ultraviolet light spectrum, invisible to us, is in the truest sense of the word a beacon for other drones intent on mating. In the ensuing couplings, the queen bee collects enough sperm for the rest of her life.

There is, however, one peaceful exception: the Cape honey bee (*Apis mellifera capensis*), which only lives in South Africa, manages completely without drones. The female workers of this species of bee lay eggs, and these hatch to produce more female workers. A puzzle that clearly represents a glaring contradiction to the logic of sexual reproduction, and which has truly baffled biologists.[4] On the other hand, it would make perfect sense if the males, having no desire to explode during insemination, had cleared off for good.

I witnessed another extraordinarily disconcerting piece of behaviour when I was a child, observing the goings-on in our home aquarium. At one time, when I had once again been banned from watching TV, I'd sit for hours in front of the tank housing our pair of cichlids of the species *Pseudocrenilabrus nicholsi* and watch with fascination as the young fry would disappear into the mouths of the parents at the first sign of danger. For me it was always a massive shock as I was terribly afraid that the small fish would simply end up getting eaten. But seconds later, they would reappear out of their parents' mouths and swim happily away. So-called mouth-brooding, which according to species is practised by females, males, or both sexes together, is a real challenge. As I later learned, the fish have to fast during the breeding period, in order that my childhood fear does not become a grisly reality. But as so often in life, this kind of love and devotion is readily exploited; the cuckoo catfish (*Synodontis multipunctatus*) hides its eggs in the clutches laid by mouth-brooding cichlids and gets them to raise its young.

But not only mouth-brooding males display this touching care for their young. Another fish, albeit one that looks more like a miniature horse, has completely usurped the position of the female. Male seahorses have their semen fertilized by the eggs of the females. The fertilized zygotes are then incubated in the stomach sac and the males ultimately give birth to live offspring.

Now that we have dealt with the broad spectrum of sexual practices such as sexual and non-sexual reproduction, parthenogenesis and hermaphroditism, let us make for more familiar shores and turn our attention to the matter of sex toys.

2. Sex toys

Not long ago, a journalist asked me whether I'd ever heard anything about animals using dildos. In actual fact, there are figuratively speaking such things as sex toys among chimpanzees, but dildos were a new one on me. On the other hand, I didn't want to categorically deny it, since of course such things aren't entirely out of the question. So I fobbed off the journalist and went away and did a bit of picture research using the search term 'animal + dildo'. To my great surprise, it transpired that there were any number of animal lovers out there. The Internet is full of rubber animal penises, it seems. Because I am assuming that neither killer whales nor horses are ordering stuff online, it must be human interest in other species that is responsible for the demand. But I drew a blank where dildos that had been fashioned by animals themselves for self-gratification were concerned. Beyond the level of anecdotes and a few absurd pictures, there was nothing. There appears to be only a single scientific paper in which an observation of this kind was even mentioned, and that was from as long ago as 1978. This paper claimed that orangutans on Sumatra rubbed themselves on lianas and branches and also possibly inserted them into their vaginas.[5] Although it's getting close, it doesn't really qualify as a description of a sex toy or dildo. At least, not according to our human interpretation, since a dildo corresponds in the broadest sense to a tool, and there are some small but important differences to be borne in mind, as we shall see.

In any event, *frottage* on a liana counts as masturbation, which is widespread in the animal kingdom. Many animals do it, and for a good reason. Lots of creatures only mate seasonally. It is possible that these species might already have gone extinct if they did not indulge in masturbation. In contrast to the female's eggs, the male's seeds can be produced on a running basis. This has the disadvantage, though, that they swiftly age and grow sluggish. Masturbation is extremely advantageous in that it ensures a constant supply of young, fresh semen, thus increasing the chances of successful insemination on the rare occasions when real sex is available. In addition, self-gratification helps relax an animal and make it less aggressive.[6] It is known that animal sexual aggression can result in some bizarre encounters. For instance, a capuchin monkey was filmed raping a duck,[7] and a chimpanzee having oral sex with a frog.[8] Also, having booked a session of dolphin therapy, the odd swimmer has found him or herself having to serve as a sex doll for the animals. A female dolphin trainer of my acquaintance went around for several weeks with angry bruises on her thighs after one such sex attack. My tip to anyone finding themselves in such a situation is please don't make the mistake of misinterpreting the first contacts as a demonstration of sympathy. After all, the animal doesn't know you, so it's inconceivable that it's instantly taken a shine to you. When a dolphin, whether in captivity or in the wild, approaches you in this way, it does so because it can, and because you are at its mercy in the water. You should immediately attempt to put a stop to such behaviour through clear gestures and an ordered retreat. And a note to my male readers, this problem isn't confined to women, since it turns out male dolphins aren't that choosy.

Now that you are clued up about what really lies behind masturbation, you'll perhaps have a bit more sympathy with those rather stupid-looking dogs that try to hump a cushion. Moreover, remedies are at hand: in contrast to the human sex dolls in the water, some companies such as 'Hot Doll' even manufacture real sex dolls for bored pets.[9]

Here, I really must scotch a widespread preconception for which we have behavioural biologists of the twentieth century to thank. They saw sex as an urge whose sole purpose was the continuing survival of the species. That's true as far as it goes, but to conclude from this that animals were merely concerned with procreation and took no pleasure in the act itself was too shortsighted. The truth of the matter is that fun, or at least the pleasurable sensation associated with it, is the central element and that self-gratification is an important interim step towards successful reproduction. In the chapter 'The hedonistic society' and part VI, 'Sentimentality' (pp. 105–7 and 269–82), we will come back to look at the mechanisms behind this, but one thing should be noted here: the pleasurable feelings that make sex appealing to us generate a high degree of inner motivation to repeat the action. In all likelihood, we share this motivation with almost all vertebrates that practise internal fertilization. At the end of the day, you have to allow your partner to get very intimate with you, and you even have to accept the risk of disease being transmitted via bodily fluids. Besides, for the duration of the sex act, both the male and female are practically defenceless against attackers. You only take risks like that if you are sufficiently motivated to do so. But what is it that motivates us? Ultimately, it is a cocktail of released hormones and neurotransmitters that allows us to ignore all the risks. All mammals operate on this level of behavioural control, and presumably birds too. Thus, we can safely assume that a hippo, a whale or a pig experiences pretty much the same feelings as we do during sex.

But feeling is just one aspect. The more complex a brain is and the more diverse the stimuli and pieces of information that it is capable of processing, the more actively and dominantly it can intervene in hormone-led processes. Our poor drones do not have this choice. Even if the appearance of the queen bee exerts no influence on them, they are helplessly at the mercy of her scent.

Scent is just as important for us humans, but thanks to our complex brains, we have the capacity to disregard it. For instance, the attractiveness of a partner interests us. But that is a product of our culture. Rubens' voluptuous beauties wouldn't have a ghost of a chance in today's modelling market. To date, we only know of a few species in which culture has a bearing on their sex lives. In part III, 'Unknown Cultures' (pp. 39–102), we will learn, for example, that orcas have a culture that goes back several hundreds of thousands of years and prohibits them from having sex with certain other groups of orcas, and that humpback whales are just as swayed by the dictates of fashion as we are. Even though we haven't got anywhere with the dildo, might there perhaps be other examples of sex toys in the animal kingdom? The answer is yes. In a review article on technology use among primates, the Cambridge professor William McGrew drew attention to a very special form of tools used for sex.[10] In an interview with the *New York Times* on the piece,[11] he cited a discovery from the year 1980 as an example.[12] There, he described the following situation: male chimpanzees are sitting around with their legs akimbo and with erect penises, and rustling leaves…

So what, you're thinking – where do sex and sex toys come into this? It all depends on how you look at it. To start with, we need to clarify exactly what a tool is. A tool is any object that is not part of the body and whose purpose is to achieve a particular end.[13] Then we should take a closer look at the leaves themselves. 'Leaf-clipping'[14] – in other words, the crumpling and noisy crumbling of dried leaves – is a gesture that chimpanzees employ to get attention. It's the same principle as if we were walking through a wood and were suddenly made aware of a bird in the undergrowth by the rustling sound of leaves. We would prick up our ears and look over at the bush from where the noise came. The female that the chimps are trying to interest reacts in exactly the same way. And what she sees stimulates a response in her. The proffered and primed instrument of pleasure between the legs of the male spells things out loud and clear. But there's no need to be envious – this strategy rarely works instantaneously, and often the poor male with his engorged member is forced to keep rustling leaves for quite a long while before the female finally approaches him lustfully and presents her bottom to him. In this moment, we have all we need in order to talk about a tool. An object that isn't part of the body and that is used to attain a particular goal.

Incidentally, male chimpanzees in the Taï National Park in Ivory Coast signal their sexual urges by cracking their finger joints ('knuckle-knock').[15] Yet if we were to bring together different chimpanzee groups that do this, things wouldn't necessarily turn X-rated: among the chimps that live in Bossou, the juveniles also crack their joints. But all they are signalling is that they want to play.

Knuckle-knocking cannot be counted as a sex toy (tool), because the knuckles are part of the body. But the Taï chimpanzees also engage in leaf-clipping. They do so when they want to get attention. Generally, the rustling heralds an important communication.

Besides, among the click languages that are prevalent across many parts of Africa, the clicks also sound like crackling leaves, and researchers have speculated whether this sound was an element of the first language ever spoken by humans.[16] It is a fascinating thought that the rustling and crackling noise of dried leaves might perhaps have been the first truly symbolic element in our linguistic development. When one considers this, the crunching of tasty potato crisps takes on a whole new meaning. Why not try your luck at the next party you go to, and dive into the nearest tube of Pringles. See what happens if you crunch away loudly while gazing into the eyes of the object of your desire. Perhaps you'll succeed in igniting primal, deep-seated animal urges – urges that are the subject of our next chapter, as it happens.

3. Rape

Two evolutionary biologists, Randy Thornhill and Craig Palmer, have conducted intensive research into the subject of rape in the animal kingdom. In their 2001 book *A Natural History of Rape*,[17] they explained the logic that lies behind this strategy through numerous examples. The study met with a very dusty reception from reviewers around the world. The problem was the fact that the authors used their observations from the animal kingdom to draw certain conclusions about human behaviour. The biological facts of the matter, which were presented extremely conclusively, sounded too much like an excuse and a justification for sexual violence among humans.

I'm not expecting you here to weigh up the pros and cons of this book, whose arguments are presented in a quite deliberately provocative way. But perhaps you will enjoy the following example, which demonstrates that evolution certainly has ways and means of keeping the most extreme male urges in check. It concerns an animal that we normally picture with a blood-smeared face as it feeds on disgusting carrion. Unfortunately, by repeatedly screening such images, wildlife documentaries have done nothing to garner sympathy for the spotted hyena. This is deeply unfair, as this species has a truly impressive social system (more on this in part IV, 'A Sense of Community', pp. 103–69) and have also evolved possibly the most extraordinary form of sexual intercourse among mammals. In the hierarchy of hyena society, males play a subordinate role; they are often smaller than the females and even the most low-ranking female can tell them where to get off.

However, this authority does not protect the females from rape, since the males are capable of banding together (about which more below) and committing communal rape, a tried and tested strategy. Even so, this is futile in the case of female hyenas, since they effectively wear a chastity belt. This appendage looks for all the world like the male hyena's crown jewels, dangling between the legs with a penis and scrotum; female hyenas have evolved an erectile pseudo-penis from their clitoris, while their labia form a sac filled with fatty tissue.[18]

The whole thing bears a striking resemblance to the male member. It effectively seals off the outermost vaginal opening. The only possibility for an erect penis to deliver its precious cargo is to penetrate the similarly erect pseudo-penis, which is constructed so as to form a hollow body. To do so, the male must virtually crawl beneath the female from behind, and both partners have to interlock their 'penises' in a well-coordinated manner. There is only one disadvantage to this perfect anti-rape defence: hyena cubs have to take this same route out of the mother's womb. As a result, birthing among hyenas is an excruciatingly painful business, and not infrequently the pseudo-penis is torn in the process and takes weeks to heal.

Yet the penis, or rather pseudo-penis, of female hyenas also performs another function. Mutual greeting among this species is ritualized to a certain degree. The hyenas stand side by side, head to tail, and the lower-ranking animal lifts its leg in order that the dominant individual can sniff its genitals. As a further gesture of inferiority, the penis or pseudo-penis is also erected at this moment.[19] Erectile dysfunction can be a real problem in these circumstances.

The red-sided garter snake (*Thamnophis sirtalis infernalis*), which is native to North America, has an easier time of it. Its chastity belt is formed by an extremely powerful muscle in front of the vagina that can prevent any penetration from taking place.[20] Because the female snake prefers to mate with mature males, this muscle is only deployed against immature males. It is a mystery to me quite how she manages this, as it is customary for several males to descend on a female all at once and to copulate in a 'mating ball'. Unlike most reptiles, which lay eggs, at the end of gestation this species gives birth to live young.

In actual fact, some animals also construct chastity belts out of straw and other materials. These look enchanting and in addition are attractively decorated. But, amazingly enough, these are the work of the male of the species. We'll return to this subject presently in the section 'Animal architects' (pp. 67–72).

4. Gangbangs

Perhaps you have had occasion to witness a poor duck being mobbed by a gaggle of drakes and forced to have sex. It's also not uncommon among charming little Adélie penguins for a group of sex-hungry males to lie in wait for a female as she returns to land exhausted after hunting for fish. One species of Amazon frog (*Rhinella proboscidea*) takes group sex to the extreme. The wretched female is besieged until she dies, whereupon the males dance around on her abdomen and push out her eggs, which they then proceed to fertilize externally. Zoologists refer to this as 'functional necrophilia'.[1]

Yet group sex need not be unpleasant per se, as the love life of bonobos shows us. At first sight, bonobos appear identical to chimpanzees. Indeed, strictly speaking they *are* chimpanzees, since the genus has two distinct species, the common chimpanzee and the bonobo. They are often called 'dwarf chimpanzees' – wrongly, as they are almost as large as their cousins. Moreover, both chimp species are more closely related to us than they are to orangutans, which also belong to the great ape family. One very simple distinguishing feature is the forehead. In bonobos it is perfectly parted and, in contrast to the common chimpanzee, lacks any hair. In comparison to the often aggressive chimpanzees, bonobos are regarded as the peaceful hippies of the anthropoid apes. Many people also believe that they are vegetarian, but there are exceptions to this general rule.[22] Their sex lives are almost impossibly confusing. Females have sex with females, males with males, females with males, and juveniles with adults. The techniques they use are also many and varied; in addition to the standard position ('doggy style'), there is the missionary position and others that are hard to describe, which involve hanging from trees. They kiss with their tongues, and oral sex is greatly favoured.[23] However, the most impressive distinguishing feature is their placidity, a quality that is almost certainly down to the intensity of their sex lives.

Just imagine for a moment that you're on a walk with some friends. The walk is arduous and to cap it all the friends of your friends are getting on everyone's nerves with their trivial sensitivities. Now your stomach starts rumbling too, but you're in luck, as there's a mountain refuge hut in sight already. Unfortunately, all the niggles along the way have seriously delayed you and by the time you arrive almost all the food has gone. The general mood hits rock bottom. Among bonobos, things would pan out differently. The first thing our nearest relatives would do would be to get an erection and even the women's first thought would be of sex. Regardless of any hierarchy, a bout of casual intercourse would ensue. Conscious of the burden he or she was placing on the rest of the group and responding to your clear antipathy, the person with the annoying sensitivities would approach and make unmistakable gestures indicating that he was interested in your genitals. After the ensuing hanky-panky, it would of course be hard to still be in a foul mood, and all would be forgiven.

In stark contrast to chimpanzees, where violence often flares up over food resources, bonobos share their food peacefully after their communal love-in. This mode of behaviour has been observed in the wild and zoos alike.[24] What's more, it can be shown that male bonobos seek out friendly contact with their chosen partner long before they go at it.[25] In other words, the males establish a relationship before entering into the serious business of reproduction. Also, there is an equality of the sexes among bonobos. One sex does not dominate the other; the opposite is of course the case with us and the other anthropoid ape species.

We all know how powerful sexually motivated actions can be. For no material or social gain, even humans – despite their highly developed consciousness and their ability to think strategically and to foresee the likely consequences of their actions – take the risk of raping another person. They do it alone or in a group or simply in their own imagination.

Where sex is concerned, even dolphins, which we think of as so nice, aren't exactly squeamish in their choice of methods. From our human perspective, we might even find their behaviour extremely abhorrent, for in this genus it is not unknown for a group of males to abduct a female in order to have their way with her, often for days on end. Many dolphin lovers find it difficult enough getting their heads around the idea that dolphins are predatory carnivores rather than vegetarians, so presumably they would renounce altogether their emotional attachment to this species if they knew about these gangbangs. But let us put aside for a moment any ethical considerations or prejudices, and consider the fact that in order to achieve their ends, male dolphins form one of the most complex social networks yet observed in the natural world.

On this subject, though, we should briefly widen the scope of our discussion while I tell you a little bit more about the life of dolphins. Unlike the situation in zoos and dolphinariums, where a male dolphin always dominates a group of females, in the wild the animals live apart from one another and only meet up seasonally every couple of years for mating. In most populations, therefore, sex with dolphins of the other sex is more the exception than the rule. In the normal course of events, the girls simply snuggle up with the girls, and the boys with the boys. Even so, once a year the urge to mate with a conspecific of the opposite gender grows too powerful and the young males, who usually plough together through the oceans throughout their lives in stable gay relationships as pairs or threesomes, coalesce into larger groups. As a result of this numerical superiority, they manage to disengage individual fertile females from their communal groups, abduct them for several days and finally get their chance, one after the other, of mating with them. But let's briefly examine how this situation – which has been observed in particular among the dolphin populations off Western Australia – comes about in the first place.

Much like human infants, dolphin babies come into the world relatively stupid. In principle, they have to learn everything except breathing and swimming, with the result that the youngsters are heavily dependent upon their mother for many years. Now, one needs to bear in mind that dolphins give birth to a roughly equal number of males and females; but given that the females are busy bringing up their young for several years, they are unavailable as partners for mating. Potential female sexual partners are therefore either juvenile or have just sent their offspring on their way. In this way, there are often five or six highly motivated young males vying for each potential female. In other words, conflict is inevitable. Unlike the life we lead on dry land, the open ocean is a three-dimensional space without any corners or edges or delimitations. It is therefore of necessity relatively difficult to separate an individual from its group, since there are simply too many escape routes open to a female. But if I, as a dolphin, go 'hunting' with a group of eight or fifteen others, my chances are greatly improved. So the sexually predatory males need to assemble a team on which they can rely. After all, it's not just about successfully abducting the female, but also about being sure that the other male dolphins will stand aside when it's my turn. In the event, the actual sex act is less brutal than we might perhaps have imagined in the circumstances. There's no question of pinning a partner down, and water is really very yielding. In a three-dimensional space, sex is an act in which both partners must actively participate, and where any disturbance can result in a failed attempt at coupling. An uncoordinated throng of randy 'bucks', with each individual fearing that the others will usurp him, is thus wholly out of place. I am not able or willing to say much about the female involved, except that nature does not share our notions of moral conduct. Yet the males' habit of forming groups has been comprehensively researched over a number of years, and the published findings are more than some flash-in-the-pan sensation, for this constitutes the first evidence of a third-order alliance.[76] What exactly that is, we will learn in the chapter 'Like Facebook, only different' (see pp. 137–47).

5. Hormones in the driving seat

The effect of hormones usually eludes our conscious perception and control. Often their influence can even be so strong that we find ourselves incapable of acting in the way we really ought to, despite our conscious recognition of this failing. Premenstrual syndrome (PMS) reveals what catastrophic effects fluctuating hormone levels can have on a woman's mood and on marital harmony. But even worse things can happen. Take the example of squid, which are generally very peaceable coevals. They like swimming around in groups and aggression among them is rare or even non-existent. But if males should happen to touch a freshly laid egg, they will flip out in an instant. At the root of this abrupt mood swing is a small protein with the name 'b-MSP-like pheromone'. This is produced by the females and stuck on the outer skin of her eggs. If a male comes into contact with an egg prepared in this way, his laid-back demeanour immediately evaporates.[27] The instant belligerence is astonishing, in so far as aggressive behaviour is normally only triggered by the stimuli of various sensory perceptions – you need to see, hear and in case of doubt even suffer a painful blow from an adversary to really get worked up into a fury. Over and above this, though, aggression itself is a truly complex phenomenon. Unlike an attractant, where all one needs to do is swim in the right direction, the squid must first identify their opponents and distinguish, say, between males and females before working out a complex order of battle.

Control by hormones becomes particularly clear when we consider the behaviour of the different sexes. We are all constructed according to a genetic blueprint and to begin with, we are all primarily female. Male individuals only emerge at a later stage in the development of the embryo.

Among clownfish (*Amphiprion percula*) – a species now etched on our minds as Nemo in the eponymous Disney film – the situation is reversed. These small anemone-dwelling fish are without exception all born as males and live in a state of polyandry. In other words, they are governed by a female, the ruler of the anemone domain with which the fish exist in a symbiotic relationship. But wait a minute, surely something's not right here – if only males are born, where do the females come from? If a female dies or unwisely ventures beyond her anemone home and gets eaten, the strongest male then begins to undergo a gender transformation. After barely a week, there is a new queen ruling over the anemone, and although the queen is the same fish it always was, its repertoire of behaviours has completely changed.[28] However, environmental conditions and mechanisms of so-called epigenesis may also be responsible for females changing back into males – for instance, among the tropical species the bluehead wrasse and the cuckoo wrasse, which inhabits temperate waters.

Changes in behaviour resulting from either the suppression or the additional administration of hormones can prove useful in a variety of ways not only to animals but also to humans. For example, male dolphins in a dolphinarium, which are prone to behave as male dolphins do, are given the drug Megace (a female hormone) and alter their behaviour accordingly. Why, you might ask, would anyone do that? The reasons are purely practical, for doing so makes it possible to keep a pod of dolphins in captivity in a relatively small pool without the animals going for one another. In the chapter 'Like Facebook, only different' (pp. 137–47), we will look in greater detail at this situation, which I know from first-hand experience at Nuremberg Zoo. A more direct approach is castration in domestic and farm animals, which is also designed to suppress undesirable male behaviours. Take a stallion, for instance. If it is castrated, it is turned into the proverbial gelding – obedient, less stubborn, a bit more relaxed, slower and in terms of pure musculature, weaker as well. Under the conditions in which it is kept, the animal gains some distinct advantages from this more agreeable nature, since geldings lead a social life for the most part. Many stallions are denied this, with the only conspecifics they ever come into contact with being mares that they are expected to cover.

From an evolutionary point of view, castration is of course less desirable, but under certain welfare conditions or in certain social systems a major intervention of this kind into the personality can even be beneficial for us humans. Eunuchs – that is, castrated males – were important personages in many cultures like Ancient China or the Ottoman Empire. In our Western culture, we immediately conjure up pictures of magnificent harems and a corpulent overseer swathed in lavish, colourful robes. The historical reality was quite different, however. Eunuchs were not particularly ambitious and were incapable of laying a cuckoo's egg in the ruler's nest, so to speak. They therefore represented no threat to the throne. Quite the contrary, in fact: they enjoyed the reputation of being outstanding civil servants and strategically astute military commanders.

Among our relatives the orangutans, one of several anthropoid apes from Southeast Asia, a similar situation pertains, though it is all smoke and mirrors. Orangutans have two male sexes. The smaller males look like females and are retarded in their physical development. They frequently remain at this stage throughout their lives. However, if a larger ape makes room in the hierarchy, they can change, conditioned by their hormones, and grow up to be large males with clearly gender-specific characteristics like throat sacs and cheek pads.

But prior to that stage, their condition is just so much deception, for unlike human eunuchs, the smaller orangutans are sexually mature and are merely waiting to seize the opportunity of the moment. Although they are not so desirable to the females, that does not inhibit them, when the boss orangutan's attention is elsewhere, from sowing their genes.[29]

If you castrate a male mammal, depending on its age – in other words, whether this happens before or after puberty – it will have far-reaching consequences for the development of its personality. But removing the testicles alone simply reduces the production of testosterone. On the one hand, testosterone is also produced in the adrenal gland, and on the other the subtle behavioural changes between male and female mammals are based on a complex interaction of several different hormones. From a modern perspective, the hypothesis that gender-specific behaviour is all about upbringing appears completely absurd. Innumerable experiments on humans and other mammals have demonstrated a seamless development in gender roles, and many of our supposed acquired behavioural traits also apply to our nearest relatives. A very simple experiment that, at least for me, is extremely convincing, runs as follows: imagine for a moment that you are a six-month-old child. You spend the whole day lying on your back and cannot move. Your head and body are just too heavy for your small muscles. The only thing you are capable of is waving your arms around a little, and you'd be happy to do that for the rest of your life. Whenever anything enters your field of vision, you try lunging at it with your tiny arms and hands, and there's hardly anything more captivating than a big, round, red balloon. Unfortunately for you, you are a guinea pig in a scientific experiment. The balloon is suddenly taken away, and you find yourself looking into two monitors. On one of them, a grown-up is gently holding the balloon in both hands, while on the other, the balloon is seen dancing merrily about, to and fro, because a grown-up keeps hitting it. After a while, the monitors are switched off again, and you are given back your balloon. If you are a boy, you would in all probability have gazed for far longer at the monitor where the balloon was seen dancing about, and you would now likewise try to hit the balloon with your hands. If you are a girl, on the other hand, you'd be in two minds and you'd behave first in one way and then in another.[30] Considering this fundamental difference, it presumably comes as no surprise to learn that boys prefer playing with mobile toys like cars, while girls tend to like playing with dolls. But perhaps it might come as something of a shock to learn that exactly the same happens when young rhesus monkeys are given cars and dolls to play with.[31] These experiments were of course conducted with captive animals, so one would be perfectly justified in asking how natural it is to get monkeys to play with toys. For this reason, researchers tried to discover whether any indications of different play strategies were evident in the wild as well. Lo and behold, that's just what did emerge, and it turned out that male chimpanzees tended to use sticks as weapons whereas females began interacting with them socially.[32] To anyone who is interested in such research findings, I would wholeheartedly recommend *The Blank Slate*[33] by Steven Pinker, one of the most influential thinkers of our time. Another eminent scientist who is strongly associated for me with the topic of hormones is Alan Turing. He was a homosexual in an age when this was a criminal offence, and so when he was arrested and charged he was given the choice of either going to prison or taking female hormones. He chose what he thought was the lesser of the two evils and took oestrogen. But under its influence, he lapsed into depression and ended up poisoning himself. To Alan Turing we owe not only some of the fundamental ideas that led to the construction of the first digital computer, including a groundbreaking experiment in human–machine interaction (the Turing test), but he also helped save hundreds of thousands of lives through his work in deciphering the German Enigma secret code, and so had a greater influence than perhaps any other individual on the outcome of the Second World War. But just through the administration of oestrogen, one of the most creative thinkers of the past century was turned into a suicidal depressive. Hormones truly are in the driving seat.

Another person who might help us to better understand the power of hormones is David Reimer. He and his twin brother, Brian, were born in Canada in 1965. After David's penis was accidentally severed as the result of a medical blunder in his infancy, his family decided on a radical intervention into their son's life, on the advice of a famous sexologist of the time, John Money. The two-year-old David's testicles were removed, and surgery conducted to create an artificial vagina from his scrotum. Thereafter, David was known as Brenda and was raised as a girl. Because the case was scientifically well documented and was designed to demonstrate that gender-specific modes of behaviour are acquired, it was for a long time regarded as pioneering. In 1975, the German journalist and feminist Alice Schwarzer cited the case of David Reimer in her book *The Little Difference and Its Huge Consequences*, and used it to corroborate the then-modish hypothesis that men and women are identical in everything but the ability to give birth. But as we have seen above, the biological reality is somewhat different, and the experiment with the Canadian twin brothers was also fated to end badly. In 1980, Brenda found out about her birth as a boy and decided to have a sex-change operation. He subsequently got married, adopted his wife's three children from a previous marriage, separated, and ultimately committed suicide.[34]

However, there are not only gender-specific hormones. Among certain mammals, including human beings, the relationship between individuals is controlled hormonally. It may sound unromantic, but even feelings like love[35] can be explained in terms of the effects of hormones, such as oxytocin.

At the time I was working on my doctoral thesis, when I was intensively engaged in studying so-called dolphin therapy, this hormone was even thought to be responsible for the supposed success of this animal-based form of therapy. Oxytocin, you see, promotes attachment and feelings of trust.[36]

At this point, I should whisk you off briefly to North America. This is the home of the prairie vole (*Microtus ochrogaster*), widely regarded as a pest. These comical little animals have gained a certain fame for the fact that they spend fully two days having sex. In turn, this drew the attention of some researchers who were interested in attachment behaviour. Sex for more than forty hours – surely that only happens when you're high on drugs? In addition, one should be aware that the majority of typical laboratory research animals, namely mice and rats, display no attachment to particular individuals and do not lead a monogamous existence but instead live in larger social communities. Perhaps it has struck you before now that most mammals, unlike the majority of birds, are not monogamous. To be precise, since the invention of breastfeeding during the rearing of offspring, we men are rendered redundant. Monogamy among mammals only appears to make sense if, through his faithfulness, a male can prevent a rival from killing the offspring in order to hasten the female's readiness to conceive once more.[37] Things are quite different in the case of birds, since they take equal turns in sitting on the eggs and in fetching food for their chicks, and as a result of this, a firm attachment between pairs is extremely common, at least during the breeding season.

It was therefore by no means easy for researchers to find a species – among the thousands of existing kinds of mammal – that practises monogamy and could possibly serve as a model organism for existence within a lifelong relationship. Yet in contrast to the closely related montane vole (*Microtus montanus*), our prairie vole is indeed monogamous.

If, as a researcher, one is interested in the effect of a substance, the tried and tested experimental method is to chemically inhibit it and observe what happens. Bingo! – when oxytocin was inhibited, our voles started to behave like their relatives in the mountains, and lifelong partnerships went out of the window. In one study only recently published, it was even shown that oxytocin was not only responsible for forging protracted, monogamous partnerships; it also stimulated behaviour that promoted the formation of a partnership, with one rodent clearly consoling the other when it was out of sorts.[38]

Thank goodness we humans, with our faculty of reason, aren't so hopelessly prey to such feelings, you might well be thinking. But your confidence would be misplaced, for even the cool calculators of the world's stock exchanges are guided by oxytocin. Test subjects who had inhaled a spray laced with oxytocin were more ready to extend credit than those who had been given a placebo aerosol.[39] Naturally, the granting of credit has nothing to do with love. But just ask yourself this: what is it that changes love to hatred? Correct, jealousy, which destroys the basis of love. Without a foundation of trust, every relationship is bound to fail. A quick huff at the oxytocin spray would also be effective in these circumstances. Oxytocin engenders a feeling of trust, which is the real social glue. That is true even of fish, which are susceptible to the oxytocin derivative isotocin.[40]

Oxytocin even works between different species. If dogs are given a shot of oxytocin by means of a nasal spray, they are better at walking to heel and are less aggressive.[41] Maybe military strategists might consider the possibility of making a chemical weapon out of the stuff, so lending the phrase 'Make love, not war!' a whole new dimension.

Oxytocin was first discovered in 1906, in connection with birth and breastfeeding, and for many years now it has been the practice to administer oxytocin after a caesarean in order to stimulate lactation. Quite coincidentally, it also triggers maternal love, for without the actual act of giving birth, the extremely powerful emission of oxytocin does not take place, nor does the almost painful attachment between mother and child come about. In evolutionary terms, this strong attachment is of great advantage, since it inclines the mother towards putting the welfare of her child and hence the continuation of the species above her own well-being. A very similar process takes place among our familiar dairy cows. Their milk production must likewise be stimulated anew every twelve months by giving birth and the powerful surges of oxytocin associated with it. How would the many millions of dairy cows around the world feel if their calves were taken away from them after a matter of hours or even days?

But perhaps it's wrong to use human feelings as a comparison; it may be that cows feel quite differently. Sadly, though, it's actually highly probable that all mammals feel in much the same way (see part VI, 'Sentimentality'; see pp. 269–82). Apart from that, it's a godsend for the pharmaceuticals sector, since all the psychotropic drugs and miracle pills that are supposed to put our emotional world back on an even keel are reliant upon animal testing.

6. Pheromone parties

You must have seen those colourful plastic balls they put in children's play areas. They're about six centimetres in diameter. Now imagine that you have to find the only yellow ball in an area the size of eight football pitches. That would be quite an accomplishment, but you'd be prepared to do that for the love of your life. Unfortunately, the balls not only cover the ground, but are also stacked 2,000 kilometres (1,243 miles) high – an impossible task. And yet the male silk moth manages exactly this using his fan-shaped antennae, which can even detect individual molecules.

The senses of the common garter snake (*Thamnophis sirtalis*) aren't quite so finely tuned. This species lives in North America and is perhaps the most intensively studied snake anywhere in the world. Male garter snakes also go after the scent of the females' pheromones. Yet these crafty chaps have a little trick up their sleeves; the males are also capable of producing female pheromones and can lay a false trail, which their spellbound conspecifics cannot help but follow.[42]

These kinds of pheromones are therefore attractants that are also effective over large distances. It doesn't matter which individuals they attract, just as long as a sexually mature male is among them. However, most mammals do not have to depend upon such acute senses, since they live in social groups, where other control mechanisms are important. The so-called Jacobson's organ, for instance, helps many vertebrates to identify some very specific signalling substances. According to species, these can be in the form of territorial markings, alarmins, laid trails, or expressions of ranking.

Humans do not possess a Jacobson's organ, and yet our noses still come to our aid in pharmacies when we're searching for the right deodorant to buy. Most people now use a deodorant so that they won't start to give off BO in a stressful meeting or at the end of a long, hard day at work. We make this decision quite consciously and systematically, as we have the capacity to think ahead and make deductions. The decision over which deodorant we should use to achieve this end, though, broaches another question. To our olfactory organ, deodorants that seem to bolster our own natural smell are preferable to others. Shouldn't our bodies therefore logically smell far more strongly when we apply them? Of course this isn't so, because what are actually giving off the smell are the metabolic end products of countless micro-organisms that feed on our sweat and dead skin scales. Deodorants contain antibacterial agents that restrict the growth of microbes and hence have the effect of inhibiting smells. But quite independently of this function, they also contain perfumes that we find pleasant or unpleasant, as the case may be. The fact that it is this scent that conditions our behaviour is rooted surprisingly deeply in the eternal struggle between single-cell and multicellular organisms, as well as between parasites and their hosts.

I studied marine biology in Kiel, and one day a fellow student, a woman, asked if she could have a tuft of my underarm hair, though she stressed I should on no account wash my armpit before cutting it. It's not unusual for biologists and medics to strike up odd conversations or ask strange questions, and so I acted like I wasn't remotely fazed by her request. She was taking part as a test subject in a scientific trial investigating whether smell had any influence on a person's choice of partner. I never did find out the result of the test, but the girl in question is now happily married – and not to me.

But what makes one smell more attractive than another, and how is it that some people quite literally get up our nose? Perhaps you have occasionally had the experience of finding the smell of close family members unpleasant. That's a good thing, since it is designed to inhibit incest and the spread of hereditary diseases. But the plot thickens even more than that, for what we are actually smelling is not hereditary disease but our immune system. To explain this, I must briefly digress. So that our body's immune system does not start attacking our own cells, each individual cell is marked. This marking is a particular protein that is specially devised for us and resembles those of close relatives. These special proteins are present on the surface of the cell membranes on platelets known as MHC (major histocompatibility complex) molecules. So far, we know of nine MHC genes in humans, each of which has 100 alleles (gene variants). Nature can therefore, it seems, roll the dice with a huge variety of possible outcomes, and every person has at least twelve different alleles. The various MHC molecules are then put together using these assembly instructions. Now, in every immune response a number of other factors also play an important part: antigens, antibodies, memory cells and several other components, but for the time being let us just go along with the idea that the number and variability of the MHC molecules are representative of the capacities of our immune system. In choosing a partner, the thing now is to find someone who will complement our own immune system and bring some new defence mechanisms into the relationship. In this way, nature ensures that the capabilities and the experiences of an individual immune system are spread throughout the whole population.[43] As if all this weren't complicated enough, our nose not only has to sniff out who is compatible with our immune system but also who is too dissimilar, for if that were the case, we would run the risk of passing on to our descendants an immune system that overreacts and begins to attack one's own body.[44]

Why am I telling you all this, then? The adaptive immune system

– in other words, that part of our immune system that adapts to new pathogens and to which we owe our capacity for immunity through memory cells – is something we have in common with all vertebrates. We all have the same problem. We look for a partner that will usefully complement our immune system, in order that our progeny might be even fitter than us. The most important insights into the mechanisms of partner selection do not derive from tests with students sniffing armpit hair or dirty T-shirts, but rather from sticklebacks. In a series of experiments, these little fish were placed into flowing water in which various MHC molecules had been dissolved. Female sticklebacks with a low MHC allele count swam towards males with a high MHC allele count, and vice versa.[45] In their choice of partners, 5-centimetre-long (2-inch) sticklebacks were found to behave in exactly the same way as people who attend 'pheromone parties'.[46] Just like us humans, they arrive at the crucial decision as to who to choose as a partner on the basis of biochemistry. If we defy this unconscious recommendation, it's quite possible that our chosen partner will quite literally end up getting up our nose. In this respect, a pheromone party, where people spend their time sniffing dozens of sweaty T-shirts, seems a shrewd first step.

Armed with this new knowledge, let us put ourselves into the world of a bitch that is ready to be serviced by a male dog. The female is in heat, and thanks to pheromones advertises herself to dogs within a large circumference as 'hot'. If a dashing dog should chance to come by, she has an inner motivation to sniff him. Should she fail to find his smell attractive, she will try to chase the dog off by snapping at him. This would mean that the dog would not get to do the deed. At some stage, his strenuous advances would tire him out and he'd jog on, possibly with a smell in his nose that hadn't pleased him anyhow. The bitch, on the other hand, would wait for another interested party to appear. In the case of arranged pairings of the kind favoured by owners of pedigree dogs, it's a tricky business. When pedigree dogs copulate, although all the cherished characteristics like charming ears and cute muzzles are passed on, so too are the less desirable features, such as retinal dysplasia, a hereditary eye disease in Yorkshire terriers. In the end, though, the random choice of partners has managed to outsmart nature, thus providing us with the explanation as to why pedigree dogs are more susceptible to infections than mongrels, which simply gave their masters the slip and took the choice of partners into their own hands, or rather noses.

7. BDSM

We humans are rather partial to BDSM, at least in our imaginations. How might we otherwise account for the huge success of *Fifty Shades of Grey*? At this point I should own up that for a long time I had no idea what BDSM actually meant. Of course, I was aware that it had something to do with sadomasochistic games, but I did not know what the letters actually stood for. Wikipedia informed me that it is an acronym for 'Bondage and Discipline, Dominance and Submission, and Sadism and Masochism'. Ultimately, what it boils down to is that one person puts another in a situation in which he or she is helpless and can be enslaved and subjected to pain – situations that would normally trigger a flight response. As we have already seen, animals with a complex brain are capable of suppressing certain behavioural reactions that are conveyed via hormones or neurotransmitters. The question is, is it only humans who derive pleasure from BDSM, or are there other species that also do so? By way of answering this question, some clever researchers began by considering what kind of experiment might get to the bottom of the matter.

Anyone who eats wasabi crisps for the first time knows what

physical reactions spicy food can trigger: the pain begins with a fiery feeling on your tongue, which goes down your gullet and then returns back up your spinal cord in a pulsating convulsion and finally, having arrived at the back of your neck, sends a tingling sensation through every hair root.

It is therefore hardly surprising that the aforementioned researchers lighted upon chilli peppers for their experiment. While these do not trigger a flight response, they can make people's food completely unpalatable and elicit a reaction of extreme abhorrence, at least until people start to enjoy the kick the chillies deliver. They tried their luck with the classic laboratory animal, the rat. The researchers bent over backwards in their attempts to get the rats to eat chillies. They slowly increased the concentration, fed juveniles with it, mixed chillies into the rats' favourite foods, and so on – but all to no avail. Only after their taste buds had been chemically destroyed could the rats be persuaded to eat food with chilli.[47] That is surprising in as much as there is nothing harmful about chilli; it just tastes very hot indeed. In contrast to rats and several other animal species tested, lots of people enjoy this taste, precisely because it is so hot. Chilli is not an isolated case either; I can vividly recall how I gagged when I drank my first glass of beer. Everything in me recoiled from this putrid, yeasty taste. I could expand this list at will: whisky, big dippers, horror films, and so on and so forth. The perversity of people's penchant for such experiences has occupied researchers for decades.[48] The scientific literature on the subject is immense, but one common denominator to emerge was people's willingness to take risks. Nowadays, this topic is being tackled in a more focused way. The desire to take risks does not necessarily go hand in hand with a love of hot food. Rather, investigators have found that people who prefer spicy food are into as many and as intense sensations as they can possibly experience.[49] Which brings us back to sex and its many variants.

I would love to be able to present you here with a list of animals that share humans' propensity to override the flight reflex and willingly subject themselves to torture, exploitation, perversions, humiliations and insults, but this does not appear to exist. Animals simply don't engage in such games. I'm happy to be proved wrong, but as things stand we seem to have hit upon the human species' unique distinguishing feature.

III

Unknown cultures

I ADMIT IT, over the past four years I have turned into something of a curmudgeon about culture. I can still recall how, before our twins were born, some of our friends urged us to get out and enjoy ourselves while we still had the chance, do something nice, maybe go to the theatre, or really paint the town red. Our friends were right. In the first year after they were born we didn't get out at all, in the second we slowly returned to our senses, and from the third year onwards we were able, at least in theory, to think about what we might do if either set of grandparents were willing to babysit. But although I'd come to regard myself as a cultural clod who didn't even seem to have any time to watch television, in actual fact I was really quite active culturally, given that culture means much more than just going to the theatre. We are all formed by our culture, and following this pattern I have also ultimately been responsible for shaping my children's culture. In doing so, I reckon I must actually have been more culturally active at that time than ever before.

There is a degree of calculation in my placing of the chapter on culture after the one on sex, since both mechanisms are responsible for the living world as we know it. Without a doubt, sex is a great thing, and was possibly the greatest invention of biology right at the beginning of the evolutionary process. However, sex has one major disadvantage: it takes an age! Of course, the actual sex act is over and done with in a trice, but the point of the whole enterprise only unfolds after several generations. So, if our tardigrade wants to deviate from his customary sexual practices, then a behavioural change has to take place from generation to generation and be reflected in the genetic code.

Behavioural change can be achieved far more quickly, though, through culture.

Culture in the animal kingdom? Surely he's exaggerating, I hear you say. Not a bit of it. Just wait and see.

Genetic selection is a process in which adaptation to certain environmental conditions and circumstances occurs over a number of generations. Cultural changes that might incidentally also influence evolution happen much more rapidly and form the basis for the human population explosion. If we cast an eye at the relevant encyclopedias, we see that they describe culture as something that stands in opposition to nature. In this, they are following a more than 200-year-old definition by the English anthropologist E. B. Tylor.[1] Even though the concept and the content of what we call culture has changed over time and across different 'civilizations', the common denominator nonetheless remains broadly the same: culture is anything that we humans do. We go to the theatre, listen to classical music or view arts programmes on the television. That all makes us feel culturally active, and prompts us to work on our cultured side. And it's precisely that last point – ongoing education or learning from others in general – that forms the heart of a cultural tradition. A particular behaviour or a tradition can therefore only gain a firm and lasting footing if the relevant information is transmitted from one individual to another.

For this reason, in recent years many biologists have begun to speculate whether such a thing as culture in a wider sense might exist, namely one that is not solely related to humans. In one of the first articles to deal with the topic of 'culture among whales and dolphins', the researchers list no fewer than sixteen different definitions of culture.[2] What is probably the best definition to date runs as follows: culture is information or behaviour acquired by means of social learning from other conspecifics.[3] This behaviour is contrasted with other expressions of behaviour. What is meant by this is animal actions that have either been inherited or which represent an adaptation to the environment.[4] For the vast majority of animals, inherited traits and environmental conditions constitute the key influence and impulsion for a particular behaviour. Accordingly, we are justified in designating certain behaviours as culture only in a very few instances, and even then with the utmost caution. The only problem is, you can't tell from a behaviour whether it is a product of culture or not. And anyhow, animal behaviour is commonly held to be based solely on instinct. When saying this, we mean some inner (intrinsic) motivation that makes an animal act in this or that way. And yet, to be perfectly honest, we don't really know what we mean by it. Such an observation does animals an injustice and at the same time prevents us from gaining a better understanding of our environment. And so we need to observe and analyse the life of animals with the forensic instinct of a detective. And who could be better suited to assisting us in our hunt for true animal culture than the world-famous Sherlock Holmes?

8. What would Sherlock Holmes' verdict have been?

The character of Sherlock Holmes, created by the British writer Sir Arthur Conan Doyle, is not only the father of all fictional detectives but also a manifestation of an intellectual revolution that took place in the late nineteenth century. Sherlock Holmes is pretty much the archetype of the logically thinking scientist who can explain the world on the basis of precise observations. His premise when making decisions is straightforward and consistent: 'When you have eliminated the impossible, whatever remains, however improbable, must be the truth!' Nowadays, the simple term for this kind of modus operandi is 'method of exclusion'. In our case, this means: if all other causes for a particular behaviour can be ruled out and the only permissible explanation is cultural exchange, then it must be a question of cultures.

In the introduction to their book *The Question of Animal Culture*, the renowned scientists Kevin Laland and Bennett Galef cite a string of incredible examples of animal behaviour. They recount stories of chimpanzees eating with chopsticks or using stones to prepare their food, of orangutans playing with sex toys and home-made dolls and of capuchin monkeys kissing each other's hands.[5] But are these instances examples of a cultural evolution? Opinions on this can sometimes be highly divergent, and many researchers voice the criticism that others often argue prematurely for a cultural explanation before having safely eliminated all other possible causes.[6] Somewhat provocatively, the critics speak in terms of an ethnographic viewpoint and accuse the researchers in question of a degree of bias and a tendency to anthropomorphize in their consideration of the issues. Naturally, the researchers reject this criticism and, fully in the spirit of Sherlock Holmes, maintain that all they are doing is correctly applying the 'method of exclusion'.[7] The naysayers substantiate their scepticism by asserting that no one can ever be sure that they have taken account of all possible influences, perhaps even some that were hitherto unknown. In a sense, of course, they are right to say this. It appears to be a glitch inherent in the system that any problem solved by science frequently throws up a plethora of further questions. We can only explain the world to the extent that we currently understand it and we should grasp fearlessly at anything that helps us to do so, even though we can never be completely certain that we have drawn the right conclusions. So, let us leave the quarrels of academics to one side and elucidate instead some basic principles before we turn to actual observations.

Our cultural heritage distinguishes us from others. For example, in my home town there is a marked difference between people who grew up in the north of the city and those from the south. Our neighbours, the French, are somehow different, not to mention the Muslims. Although we have been differently socialized, we all belong to a single biological species and can call ourselves humans. Basically speaking, culture only exists within a species, though this is not to preclude the possibility of there also being a cultural transfer from one species to another. For the present, however, let us concentrate on the spectrum of possible behavioural mechanisms within a species. Representatives of one and the same species can sometimes either look different and behave in the same way or behave differently and look identical. Only a genetic analysis provides absolute certainty, a strategy that also enables us to distinguish between different populations within a species. Differences within a species often lead to a division into two distinct species. This is the basis of evolution. Biologists refer to the process as allopatric speciation, when a species splits into two as a result of changing environmental circumstances. The influence therefore comes from an external source. By contrast, sympatric speciation is when the division occurs from within. An example of this would be the medium ground finches (*Geospiza fortis*),[8] one of fourteen different species of finch discovered on the Galápagos Islands in 1835 by Charles Darwin. Among the males of this species, there are two differently sized types of beak. Depending on which form of beak they have, the calls of the two groups of males sound different, with some females preferring one song and some the other.

If one did not know about the effect of the beak shapes, one might very well be tempted to regard the behaviour of the two different groups as a distinct culture. One might assume from this that each individual group was following in a particular cultural tradition. This behaviour might, for instance, be a particular song that is only transmitted within a particular cultural group. Yet in actual fact we are in the process of observing the division of a species into two distinct species on the basis of a small morphological difference. This observation is of enormous significance to researchers into evolution, since to all intents and purposes it allows us a ringside seat at the emergence of new species. However, that has nothing to do with cultural differentiation.

But apart from different calls, what other behaviour patterns are there? Let us consider reflexes, for instance: these are very simple, genetically predetermined behaviours such as the sucking reflex of a newborn mammal. One characteristic feature of this is the fact that the behaviour remains unchanged over very long periods and often extends across several species.

Very few people are aware that carrying babies also triggers a reaction in the infants. Some people might believe that this must be a deeply human behaviour produced by feelings of love and devotion, but baby mice react in exactly the same way. In an experiment, drugs were administered to dull their senses when separated from their mothers, although the mothers were free to fetch back their young. Stupidly, the baby mice wriggled like crazy in the process, and consequently their mothers took far longer to carry out their rescue mission.[9] In the normal course of events – that is, without any drug intervention – on being picked up the baby animals immediately lapse into the so-called transport immobility. This reflex can be triggered in the great majority of mammals and allows the parents to get their young quickly to safety. All mothers and fathers who are frustrated that their babies, which were so peaceful and calm when they were being carried around, immediately start crying again once they are put to bed should rest assured that this is quite normal, since the transport immobility reflex is no longer being triggered. But forget trying to pick the baby up again; the root cause of the problem – be it hunger, thirst, or a full nappy – won't be remedied that way. When all's said and done, we are all just small animals.

Then there is behaviour that is genetically predisposed. This is a learned response, which is often species-specific. A typical feature of this is that it can be acquired especially easily by the individual. A prime example would perhaps be a talent for music: if we look at two children trying to learn the guitar, it is highly probable that a child from a musical family, where music has played an important role for generations, will learn to play the instrument better and more readily than one for whose forebears music did not play any significant part. Yet if the children are not presented with instruments, the differing genetic predisposition will not come to light in the first place. For the observation of behaviour in animals, this means that certain behavioural patterns only manifest themselves under particular environmental conditions.

A further consideration is the way in which things are learned, or in how a behaviour is acquired. If an animal displays a new behaviour pattern, it is perfectly possible that it taught itself this mode of conduct or acquired it through experience. In this case, too, this would not amount to a cultural property. But if this self-taught and not genetically predisposed behaviour is seen and copied by other animals, then it is indeed an instance of a cultural transmission of information.

You can therefore see how complicated it can be to prove incon-trovertibly that a behaviour is culturally acquired. Unequivocal observations in the wild are extremely rare and experiments on this subject highly complicated.

Behavioural biologists are keen to find the most simple systems possible for their experiments. Naturally, they do this not because they are lazy, but because a simple system is easier to test, so reducing the incidence of any potential sources of error. In one such relatively simple experiment, scientists were able to show that a culture existed among coral-dwelling fishes.[10]

When animals mate, this activity demands their full attention. As a result, they do not have much time or energy to spare to watch out for adversaries. Accordingly, biologists assume that mating grounds must have particular characteristics. Humpback whales from the northern Pacific, for example, swim to the calm, protected waters of the Hawaiian Islands in order to mate and rear their young. In principle, therefore, it ought to be possible to describe the conditions required for a mating ground. In this way, one could reasonably predict that a certain area would be suitable as a mating haunt or 'lek'. In order to prove that experimentally, one would thus need to take a relevant population and transfer it to a different environment. One could then observe whether the predicted areas were indeed used in preference to others. Of course, not only ethical but also practical considerations preclude the possibility of simply transposing the humpback whale population from the Pacific to the Atlantic, just in order to see whether they sought out similar bays in the Caribbean. In contrast, relatively few difficulties arise when experimenting with fish. So, an entire local population of a small coral-dwelling fish called the bluehead wrasse (*Thalassoma bifasciatum*) was caught and released into another area. It should also be noted here that the population in question had already been under observation for more than four years and over several generations. The researchers therefore knew that the animals sought out a particular area time and again for breeding.

If the environmental conditions were decisive in the choice of breeding ground, then logically the wrasses ought to search out a similar area in their new environment in which to breed.

Bluehead wrasses are not especially solicitous in the matter of breeding and raising their young. To be precise, there is only one requirement, namely to be in the right place at the right time. Males and females then need to release their eggs and semen into the water at the same time. And that's that; Mother Nature does the rest. This makes it all the more important that the coordination between the two sexes runs like clockwork. After the population had been transferred, they did indeed find and use a new breeding ground that remained stable over several generations. Even so, the researchers came to the conclusion that the choice of area and the constancy of its use had nothing to do with environmental conditions. It appeared rather to become a kind of convention among the fish; this was then passed on as information from one generation to the next. This example shows that very simple cultural information can also be transmitted by relatively less highly developed animal species. In this case, the fish fry simply swam around with their parents and in turn their offspring likewise took their cue from them. A simple and efficient system! Culture as such is therefore emphatically not some outstanding cognitive achievement. Learning from others is nothing more than a thoroughly practical mechanism. Yet what is passed down from one generation to the next can vary enormously in its complexity. Some researchers therefore take the view that the learning performance of our little fish is not a cultural phenomenon. They talk about 'one-trick ponies' and require that the animals in question should not merely be capable of one specific feat, but should also fundamentally display the basic requirement for a culture, namely the capacity to learn and pass on other or new information.[11] The fish therefore inhabit a half-culture. Thus, as with so much in nature, the capacity for a cultivated lifestyle likewise evolved only gradually. It is even possible for an animal to have a half-self-awareness (more on this in the section 'Self-awareness'; see pp. 196–204). Accordingly, in both evolutionary and cognitive terms, it is a long way from simply swimming behind your parents to the reading and writing of books or browsing the Internet.

In order to do proper justice to this spectrum, two pioneers in the debate about culture, the researchers Andrew Whiten and Carel van Schaik, proposed a multi-stage model (the 'cultural intelligence hypothesis').[12]

Social learning through imitation can frequently be observed in the animal kingdom. Nonetheless, a distinction must be made between genetically preprogrammed learning and learning afresh. Although a chick that learns from its parents how to fly is learning socially and through imitation in the broadest sense, it is learning a genetically preprogrammed behaviour. Yet when animals are able to imitate new behaviour, then they fall into this category.

Tradition is the transfer of socially learned new behaviour (often over several generations) within a group or population. Examples of this are the dialects among orcas, starlings and mice (see 'The secret language of animals'; see pp. 74–94).

Culture is defined as the conglomerate of different traditions. A culture therefore consists of a multitude of traditions or, more accurately, of the ability to absorb new traditions into one's own culture at any time. This distinction should make clear that an individual tradition, such as we have identified in fish, does not at root mean that these animals are capable of developing a complex dynamic culture.

Cumulative culture is regarded as being the sole preserve of humans and a very few other species. Cumulative culture represents among other things the ability to divide up into groups and specialize. An example of this would be the hunting strategy adopted by chimpanzees, which involves various different roles (see the section 'Strategic thinking and creativity'; see pp. 183–90).

Only very recently, this model was refined by an interdisciplinary working group and published in the work *The Nature of Culture*.[13] The term culture was rechristened 'basic cultural capacities', and cumulative culture was subdivided into three separate capacities:

- Modular culture: this capacity is represented by the use of one tool to manufacture another.
- Composite culture: various objects (such as stone tools and wooden handles) are combined to form composites with new qualities.
- Complementary culture: in this capacity, any resulting action is based on the combination of different individual actions of various group members.

This renaming and expansion of the original categories seemed important especially to the researchers, who were attempting to gain a better understanding of human evolution. Undoubtedly, this more precise specification brought more clarity to the debate. But from my point of view, it also creates a problem: the various different aspects of cumulative culture are based almost exclusively on the use of tools. Yet we cannot exclude the possibility that other forms of cumulative culture are organized in a quite different way. It is possible that these other forms have placed greater emphasis on social and emotional aspects and observe certain rituals that are inaccessible to us. It may be the case that such cumulative cultures defy direct observation on our part and of necessity are much harder to investigate, given that there are no archaeological relics in the form of stone axes for us to study. Let us therefore concentrate instead on the cultural pyramid.

My former colleague Mike Bossley has for many years been

studying the local dolphin population off Adelaide in South Australia, and some years ago he observed a behaviour that we are only now in a position to properly understand. In the 1980s a female dolphin by the name of Billie was unfortunate enough to stray into the sluice of a harbour basin. She was caught and kept for some weeks in a dolphinarium. Luckily for her, it was decided to release her back into the wild thereafter. To the great astonishment of all those involved, though, she had brought something with her from the dolphinarium, namely a completely unnatural pattern of behaviour. If you've ever visited a dolphin show, then you'll surely have seen dolphins swimming upright using their dorsal fins for balance. This is a favourite feature of every performance, as their bodies rise up almost completely out of the water, giving the spectators an excellent view of the animals. But this is not part of the natural repertoire of behaviours for dolphins, nor indeed is there a single animal out there in the wild throughout the world that balances on its tail – except for our female dolphin, Billie. She learned this trick from watching the dolphins in the dolphinarium. What is remarkable is that she wasn't trained to perform this trick or any other. Yet perhaps she found this behaviour so spectacular that, once back in the ocean, she simply tried it out for herself. This, then, is clearly an instance of social learning through imitation, in which a completely new behaviour was adopted. It is also interesting that she did not display this behaviour in the dolphinarium, meaning that she evidently had the capacity to remember the time she spent there.

We find a comparable behaviour among orangutans on Borneo. Here, a number of animals that have likewise been kept in captivity by people imitate human behaviour, in their propensity for washing socks or other cloth-like materials.[14] In this case too, it is a question of social learning through imitation, here applied to human behaviour.

There is an addendum, however, to the story of our dolphin in

Australia, for around fifteen years later my colleague noticed that other dolphins also began balancing on their tails.[15] This therefore showed us how a tradition could develop out of imitation. Along with other learned patterns of behaviour, such as characteristic calls or particular hunting strategies, this provides us with a clear example of a culture. And if we look at the distribution of different roles in certain hunting strategies, this is even a form of cumulative culture.

After having had to absorb so much theory, you deserve a few fascinating concrete examples, so let's dive in straight away with something that most people think only our own species is capable of.

9. Music and fashion

Let us now test the knowledge we've just acquired on the example of humpback whale songs; I'm betting that a good half of you have a CD of them on your shelves. In the first instance, we should note that acoustic communication is also a form of behaviour. The use of communications signals may be inherited, but can also be learned. Examples of inherited signals are dogs barking or cats miaowing, while birdsong is learned. Acoustic communication in particular presents several advantages for the research of culture: animals that can imitate acoustic occurrences are eminently suitable candidates for research into cultural accomplishment. This is simply due to the fact that acoustic signals can be easily changed and be extremely varied. Scientists refer to this property as plasticity. No specific environmental conditions are required for this, therefore, such as the presence of particular resources, for instance. Accordingly, researchers do not have to wait to observe a particular hunting strategy for a particular prey item, nor does one have to search around in order to locate a termite mound where chimpanzees are using sticks to probe for tasty ants. Furthermore, acoustic signals do not cost much energy and they can be produced at any time – ideal requirements when one wants to study the transfer of information from one animal to a whole population.

We have already known for several decades that the songs of humpback whales change over the years. Every year, the humpback whales in the North Atlantic, for example, reinvent about a third of their songs, and researchers have noticed that the repertoire of songs changes completely over a span of fifteen years.[16] But among the humpback whales off the east coast of Australia, this same changeover only takes two.[17] This period is of course far too short to allow any latitude for speculation about genetic causes. Also, the fact that all animals within the population attempt each year to include the same new changes in their repertoire is clear evidence of a behaviour learned in a social context. What we have here, then, is a classic case of a genuine cultural asset.

Encouraged by the results from the east coast of Australia, scientists also proceeded to study other populations of humpback whales in the Pacific. Their findings prompted them to talk in terms of veritable cultural waves migrating across the Pacific over several years.[18]

The researchers worked from the assumption that the new elements in the song repertoire make the 'singers' more attractive, or that quicker adoption of a new element is a sign of biological fitness. This is not unlike the dictates of fashion in the human world, for by being fashionable we too are trying to show that we are abreast of the times and at the cutting edge of modishness. Our brain fools us into believing that we find the latest pop song, the particular cut of a shirt collar or a new combination of colours more attractive and modern. Some inner impulsion that is beyond our ken causes us to regard things that chime in with the contemporary zeitgeist as better, and we buy into the project of being 'up to date', lock, stock and barrel. Presumably, though, that is precisely the biological trick, since only a person who is fit could afford to expend so much effort for nothing in the first place. Seen objectively, yesterday's whale songs or the fashions of twenty years ago are just as good as those of today. Yet following fashion works well and is good for the economy. One need only consider how many jobs this clever biological wheeze has created.

10. On good taste

We're not talking here about Swiss chocolate or Italian cinnamon ice cream, but rather about cultural taste and the often very rigid rules deriving from it, which are applied forcefully and sometimes even with violence. An acquaintance of mine is a microbiologist who works at one of the most renowned laboratories in the world; she has a doctorate and is considered an authority in our discipline. Some years ago, her parents promised her hand in marriage to the son of a couple they were friends with. Another female friend is happily married and a mother, but her marriage was only made possible by her changing religion. In both cases the older generation had an influence on the choice of partners, thus ensuring the maintenance of their own culture. And even though both examples derive from my personal experience within my own milieu in Germany, they do not form part of our Western culture. To our way of thinking, behaviour such as this appears backward, because we put a premium on the needs of the individual and are firmly of the view that individuality makes for more freedom and creativity. And yet to the representatives of both foreign cultures in Germany cited in my examples – from India and Turkey, respectively – it might conversely appear to signal a total lack of culture if the older generation were to have absolutely no influence on the choice of partner. If orcas understood the concept of our cultures, then they would in all likelihood feel more in sympathy with the Turkish or Indian way of doing things. Is this surprising?

Along the western seaboard of Canada, there are three groups of orcas: the residents, the transients and the offshore orcas. All three groups use the same territory, at least some of the time, and it is something of a mystery that they do not mate with one another. In fact, from a genetic point of view, mating among the groups would be eminently sensible, since it would avoid inbreeding and expand the gene pool. However, genetic studies have shown that the transients split off from the residents and the offshore orcas some 700,000 years ago, whereas the residents and the offshore orcas only came to form separate populations from around 150,000 years ago.[19] By way of comparison, the evolution of the human brain came to an end around 200,000 years ago and has not altered in its basic structure ever since.

Even nowadays, offshores and residents do seem to mate very occasionally, notwithstanding the fact that they meet less frequently than residents and transients. Interestingly, residents and offshore orcas also share other characteristics: they do not eat any mammals but only fish, and both groups are positively 'loquacious' when hunting – in other words, acoustically active. Transients, on the other hand, are almost silent during hunts, as they do not want to be heard by their prey.

After all that we have learned here about culture, the only explanation for the well-documented communication ability of orcas must be that it is a cultural phenomenon. For this reason, and after ruling out all hitherto recognized possible influences on the different behaviours of the animals, various groups of researchers ultimately reached the conclusion that it really had to be a question of divergent cultures among the orca populations.[20] Ongoing studies go a step further and regard this example as one of the very few instances thus far observed in which a cultural difference has had an influence on genetic evolution.[21] A comparable example among us humans is the lactose intolerance to which many ethnic groups are prone.[22] As you may well know from your own experience, we humans are the only animal species that partakes of baby food in the form of milk when we are adults. Those peoples that practise intensive animal husbandry and among whom adults also ingest milk have over the course of the past 10,000 to 15,000 years developed a tolerance to lactose. By contrast, most people in southern Asia, South America and Africa have retained a lactose intolerance that is fully commensurate with the natural dietary habits of humans. A typical instance where a cultural achievement – in this case, the rearing of domestic animals – had an influence on genetic evolution. Formerly, it was assumed that the cultural influence was only sufficiently strongly pronounced among us humans for it to influence evolution.

But what might be the reason for the different populations having developed among the orcas in the first place? To be perfectly frank, we do not know why these orcas cultivate such diverse cultures. One thing is absolutely certain, though: transients and residents have wanted nothing to do with one another for hundreds of generations. The following two examples will perhaps clarify how deeply embedded this cultural legacy is among these creatures.

A predilection for certain foods is well known to be widespread within human culture. A Muslim eats no pork, a Hindu would never touch a cow, while those of us rooted in Western culture would find our stomachs turning if we were to witness a Korean chowing down on a dog leg. Yet how far would we go? Would we regard starving to death as an acceptable price to pay for upholding these preferences? A group of three transients that were captured for a dolphinarium in British Columbia in 1970 perhaps asked themselves that very same question: for no less than 75 days, they refused to eat fish until one of the animals starved. Whereupon the remaining whales decided to start accepting the food that was offered. Interestingly, these two orcas were later released back into the wild. So what did they eat there? You got it in one – they reverted to their original or traditional diet of marine mammals.[23]

To cite another example: the foundering of the oil tanker *Exxon Valdez* in Alaska in March 1989 led to one of the greatest man-made environmental catastrophes ever witnessed. This included the death of nine orcas from the pod AT1 who were living in Prince William Sound, the site of the shipwreck. The loss of these nine animals accounted for fully 41 per cent of the local population. Twenty years after the disaster, there were only two sexually mature females living in the sound. Unfortunately, there was no observational evidence to suggest that these animals mated with other orcas, and researchers are assuming that as a result of the *Exxon Valdez* catastrophe the local population will die out.[24]

Both of these examples show how deeply the cultural tradition must be rooted among these animals. Even in a life-threatening situation or when the entire population was in jeopardy, the animals hesitated to do something that would in theory have been no problem for them. The only question is: what exactly is it that prompts the animals to behave in such a way? Is there such a thing as a whale ethos, under the motto 'We won't eat smelly fish!' or 'We don't want anything to do with residents!'

Culture among orcas has also been observed in captive animals. At the Marineland dolphinarium in Niagara Falls, one orca lighted upon the idea of using bait to catch seagulls.[25] He would regurgitate scraps of fish from his last meal and lie in wait until some bold sea-gull swooped down on this new-found morsel. It did not take long for the orca's younger half-brother to notice the trick and to start hunting seagulls himself. And two adult female orcas subsequently took their cue from him. This was a clear case of cultural transference;

and here there was no possibility of interpreting the behaviour in any other way.

Maybe the following thought will cause us to reflect rather self-critically on our own culture: *Homo sapiens* – but also various extinct members of our genus, such as *Homo habilis, Homo rudolfensis, Homo erectus* and *Homo neanderthalensis* – look back on around 2 million years of use of stones as tools.[26] A significant achievement, and one of which we can be justifiably proud, but this is nothing in comparison with orca culture. Why is that? It is actually relatively difficult to maintain two distinct cultures in one and the same domain. At some stage, we humans began putting up fences. We then eagerly set about shifting these back and forth, with the result that a particular region saw this or that culture predominate at any given point in time. In the history of our species, there are very few examples of cultures that were able to retain their independence from their own territory and within another culture for any significant length of time. Examples of this might be the Sinti and the Roma peoples, over several centuries, or the Jewish people, with a tradition lasting for several thousands of years. But what is that when compared with the 700,000-year-old cultures of the orcas on the west coast of Canada?

Finally, though, I will treat you to one more anecdote on the subject of good taste, even though it's a moot point whether it's proof of a culture. If you were of the view that we humans were the only animals to enjoy putting together delectable menus or mixing elaborate cocktails, you'd be wrong. Naong, a twenty-one-year-old male orangutan, was as well versed in the tasty composition of different flavours as the students who were the subjects of a comparative study.[27] However, we will gain a better understanding of this incredible achievement presently, when we come to treat the topic of memory, abstract thought and creativity in part V, 'On Thought' (p. 169).

11. Patent office or open-source?

When I was a boy, one of my favourite films was an Italian comedy called *Mr. Robinson*, a parody of *Robinson Crusoe*. It tells the story of a businessman who has been spoiled rotten by modern technology and who finds himself stranded on a tropical island. He is hungry and thirsty and all around him are masses of coconuts. His countless painful attempts to crack them open form the film's running gag. In our modern-day cumulative culture, this naturally poses no problem. Before we try to break open a coconut, we visit YouTube, where we are bombarded with dozens of different solutions. And we'll be offered a special tool for opening coconuts. It looks for all the world like a screwdriver that ought to cost about £1, but what they're actually charging you is £15, and you can't even use it to unscrew things.

This simple example shows how vital it is to share information and cooperate. We humans have become past masters at this discipline. Here's an example you'll all be familiar with. You're browsing the Web, wanting to send off an online order or to log in to a site and you're asked to fill in a CAPTCHA. That's one of those hard-to-read sequences of letters and numbers that are often displayed in a distorted fashion.

These little pictures are currently illegible by computers, but the human brain can manage this task with ease. On average, you need about ten seconds to enter the information. Worldwide, this happens around 2 million times a day. Adding all this time together means that a total of 240 years of people's lives go down the drain. This wasted time can be laid at the door of Luis von Ahn, the man who devised this system. Luis von Ahn is a clever man who is not indifferent to this problem, and so he added a little something to his invention.[28] Now there are a number of websites that ask you to type in two words instead. Like me, you may well have been irritated by this development and wondered how long it would be before you'd be required to enter a whole sentence into the fields. Far from it, though. It doesn't take much more time to enter the second word, and the operation has nothing to do with enhanced security. The second word you enter will have come from an old book that has been digitized but could not be read properly by computer programmes. In other words, when you type in a double CAPTCHA you're actually participating in one of the most important projects ever devised by humans. You are helping to rescue old knowledge for the computer age. Thus, we have evolved tools with whose assistance we can collaborate with the rest of humanity, and moreover do so without realizing it.

This project goes right to the very roots of our information management. Before we discovered that knowledge and information equated to power, we lived in an open-source community, in which insights were swiftly shared. The exchange of information with other Stone Age communities created a situation in which cultural accomplishments could accumulate, hence the term 'cumulative culture'.

The invention of patent offices and the crazy idea of keeping information secret are therefore precisely the opposite of what has enabled our society to grow. Fortunately, Wikipedia, Wikileaks and the whole open-source revolution of recent times run counter to this trend. Unfortunately, our closest relatives – the three anthropoid ape species: chimpanzees, gorillas and orangutans – no longer have the ability to develop a cumulative culture of this kind. How close they came to taking this important evolutionary step is indicated by the following example of a material culture that is at least 4,000 years old.

*

'Chimpanzee Stone Age'

Our closest relatives, the chimpanzees, also appreciate the problem faced by 'Mr. Robinson': they would dearly love to crack nuts, but are incapable of opening them through sheer physical force alone. The coula nuts they are after look a bit like walnuts, a resemblance that has earned them the alternative name of African walnuts. Yet the two are not related. The shells of the coula nut are so hard that even many seedlings fail to break through their own protective wall. There's no chance of opening them without a tool to hand. But almost as if our clever relatives had watched YouTube videos, they use hammers to crack open the nuts. It is intriguing that different groups, despite having the same living conditions, use different objects as hammers. Some employ both wooden and stone hammers, while others use exclusively hammers made of stone.[29] Naturally, wood and stone are readily available raw materials in both territories. Perhaps at this point you will recall our introduction to the subject of culture and researchers' efforts to draw distinctions between various degrees of cumulative culture. One important criterion was that tools should be made up of various components. But that would be a step too far in this instance; chimpanzees aren't capable of fixing stones to pieces of wood in order to fabricate a genuine hammer. They do, however, have another impressive trick up their sleeves. Anyone who has ever been in the jungle before will know the sensation of the ground feeling hollow, even spongy and yielding, beneath your feet with every step you take. The topmost layers of soil consist of decayed plant material and countless smaller and larger crawling animals that live off this detritus. If you take a hammer and bring it down on a hard nut placed on the ground, both it and probably the hammer too will disappear into the soft rainforest floor. The trick with the hammer only works on a hard base, and because there are no tiled floors or marble kitchen countertops in the jungle, our relatives use tree roots as a kind of anvil. In this way, they fulfil the criterion of a composite tool, since they would be unable to achieve the desired end with either the anvil or the hammer alone.

Perhaps you are asking yourself now whether our relatives in Africa really have watched YouTube or perhaps learned the trick by observing us humans. Answering this question gave the scientists a golden opportunity, since at last they were able to use the methods of archaeology to investigate culture among animals. And so they set about demarcating a small area and started to excavate for old stone hammers. The result was that some plant remains clinging to the hammers were found to be more than 4,300 years old; at that time, the Taï region of Ivory Coast, where the study was conducted, had no human inhabitants. Little wonder, then, that the researchers thereafter began talking about a 'Chimpanzee Stone Age'.[30] In addition, a number of experimental studies of captive animals on the spread of various feeding strategies have clearly shown that chimpanzees are capable of cultural achievements.[31]

Some caution is in order here, however, as one of the most famous postulated instances of a material culture turned out to be fallacious and nowadays serves as an example of how quickly we can go astray. The phenomenon in question is so called ant-dipping.[32] This involves chimpanzees poking around in anthills with sticks they have fashioned specially for the purpose. The enraged insects swarm on to the supposed intruder and sink their mandibles into the stick. At this point they are abruptly pulled out of the anthill, and just a second later they are licked off the stick and consumed by our smart chimpanzee. The ants eaten in this way are presumably an important source of nutrition – though clearly not an essential one, since not all chimpanzees in Africa eat ants.[33] Then the researchers noticed that different groups of chimps used sticks of differing lengths. It is therefore hardly surprising that the scientists immediately began to speculate that these different lengths of stick must represent a cultural tradition. On closer inspection, however, it turned out that the length of the sticks correlated to the relative aggression and speed of different ant colonies. Long sticks were used on aggressive ants and shorter ones on less aggressive insects.[34] As such, the supposed instance of culture was revealed to be nothing more than simple adaptation to given circumstances.

Even so, our close relatives observe others far less than one might have expected. Presumably most chimps learn tool use through trying it out for themselves rather than from one another. In any event, they are past masters at experimentation, and a study showed that they behave just like human children in this regard. The researchers set a group of two- to four-year-old children and anthropoid apes some fiddly tasks, which could only be successfully completed with the spontaneous aid of tools. It transpired that there were hardly any differences in the way the problems were solved. The researchers concluded from this that the cognitive bases for spontaneous tool usage are comparable across all anthropoid ape species, including us humans.[35]

Heavy nomads and light nutcrackers

Elephants likewise live within a culture, and what's more in one that we must define as a nomadic culture. Their migratory routes, and the manner in which various natural resources such as water and food are acquired, follow certain traditions.[36] African elephants have been found to actively use a living space of some 10,000 square kilometres (3,860 square miles) – an area equivalent to almost half the size of Wales. Having inhabited this region for many generations, they know it intimately, probably far better than you or I know our home towns.

These elephant migrations have nothing to do with a predator roaming through their territory or with the seasonal movements of birds or whales. Elephants' migrations follow their own patterns that cannot be explained by external influences such as the quality or availability of food, the presence of predators, and so on. They have even cultivated their own landscape, for in some places there are special 'elephant baths' – small watering holes whose fringes they regularly keep clear of vegetation.

The elephants have learned the peculiarities of their territory from their parents and grandparents, and now roam through the terrain in much the same way as our ancestors must once have done, or as nomadic peoples still do. However, on their wanderings they encounter one annoying problem: even these pachyderms are not immune from being fed upon by blood-sucking insects, and to mitigate this they construct 'fly swatters'[37] from tree branches. Yet it would probably be unjustified to talk of a culture where this form of tool manufacture is concerned, since there is too great a probability that the animals simply devise this trick of their own accord. After all, the way they get food largely involves tearing off leaves and shrubbery. It's amusing to observe, though, that when they are not using their swatters, they stick them behind their ears like a carpenter might do with his pencil.

Of a quite different nature, however, are the problems encountered by some bird species that, like our chimpanzees, want to get at the contents of certain tasty nuts. They too use tools – in their case, cars,[38] – and in the process deploy just as clever a trick as Luis von Ahn's with the digitization of old books.

Several corvid species fly with nuts or mussels in their beaks and deliberately drop them in the hope that they will crack open below and release their delectable contents. Of course, every flight entails expending a certain amount of energy, and it's all the more frustrating when one of the hard shells refuses to break open. Some crows must have realized that roads are really hard surfaces and that even the hardest nuts smash when a car runs over them. But roads are also very dangerous places, and there is a high risk of paying for the desired morsel of food with your life. At some stage, a crow must have hit upon the idea of doing this at a crossroads where, as if by magic, the noisy, stinking vehicles miraculously came to a halt when a little red light came on. When the amber light lit up, you needed to get yourself off the road pretty smartish. Another advantage of this method was that you didn't even have to fly up very high. Without a doubt, this is an impressive achievement, even if it came about by chance rather than through any strategic planning. But what then subsequently transpired was truly astonishing. The birds learned the trick by watching one another and now, thirty years after it was first observed there, this method is widespread throughout Japan. Of necessity, the transmission of this technique cannot be explained genetically over so short a time span, nor is it likely that the crows accidentally lighted upon the same idea solely in Japan, while crows in the rest of the world remained ignorant of it. According to all we have learned thus far, therefore, we are left with the emergence of a tradition and hence a cultural explanation.

Recently, a researcher from Oxford by the name of Dr Lucy Aplin even managed to produce experimental evidence of cultures among birds. She trained a number of wild-caught great tits to operate a simple mechanism. The apparatus in question was a small box that released food after being pecked at for a short while. These boxes were then set up in the wild, and the birds that had been trapped for the experiment were released. It took no more than a week for most birds in the wild to learn the new behaviour. This unequivocal proof of cultural mechanisms was so impressive that the findings were published in *Nature*, the world's most prestigious scientific journal.[39]

'Spongers' and 'shellers': animal lifestyles

The term 'lifestyle' describes a characteristic manner in which a person leads their life. Might it be possible for animals also to cultivate particular lifestyles? Could it be the case that different hunting strategies constitute one and the same kind of lifestyle? Probably not, for lifestyles are characterized by people who share the same environmental living conditions opting for different ways of leading their lives. Eco-freaks, esoterics, yuppies, gym-bunnies, Mr and Mrs Average, and squares – to name just a few.

This situation is hard to conceive of in nature, although the foundations have already been laid (as we shall see in the chapter 'Who am I, and who exactly are you?', pp. 196–230). The most likely potential matches are the differing ways of life among the aforementioned orca populations. Yet their lifestyle is culturally so firmly established that a voluntary decision seems improbable. Nevertheless, we do find an impressive example of lifestyles among dolphins. Imagine for a moment that you have to clean a kitchen work surface or fetch wood for the stove or change a tyre on the car. What group of people do you belong to – those who wear gloves to do these tasks or those who just get the job done as quickly as possible? In my experience, it is barely possible to persuade a glove-wearer to perform these tasks without them. He or she would be perfectly capable of doing so, but only does so very reluctantly; it's just not their style. This might well also be the case among dolphins.

Throughout the world, dolphins that live in shallow waters display a very typical behaviour. Standing on their heads, they love burrowing down into the sandy or muddy seabed. While doing this, they send out pulses of ultrasound into the substrate, in the hope of catching any small fish hiding there. Off the coast of Western Australia, some of these burrowing dolphins take the precaution of equipping themselves with gloves. Of course, having no hands they don't *literally* wear gloves, but instead hold a sponge at the tip of their rostrums to protect themselves from sharp-edged mussel shells and stones while digging.[40]

The researchers dubbed them 'spongers' and with this form of tool use they have even set a record among animals of all kinds. Normally animals use tools only very infrequently, and the length of use scarcely amounts to more than 10 per cent of any one day. But among the 'spongers' it has reached 17 per cent. And there are other puzzles to ponder. It remains a complete mystery why only around 50 per cent of the total of around 2,000 individual animals in the Western Australian dolphin population use sponges.[41] Even more so when they are all related. But it is also a fact that not all animals in this lineage use sponges, and those that do are pre-dominantly females. The animals therefore have a free choice, which is why researchers have also designated this behaviour as a lifestyle.

However, the plot now thickens, for in contrast to the wide-spread free exchange of information in an open-source society, these animals appear to want to keep this aspect of their behaviour a secret. Two factors should be borne in mind here. First, when the dolphins go hunting with sponges, they are mostly on their own, and only in 6 per cent of the sightings were 'non-spongers' in the vicinity. Although it would only take a few seconds for an individual to learn the actual technique through observation, nonetheless the associated hunting strategy seems to be relatively complicated, for young animals learn it only very late, after they have already become experienced predators. It almost appears as though the animals want to guard their secret and only pass it on to their relatives. Have they perhaps realized the added value of their hunting technique? That would certainly be a plausible explanation for the technique not having become more widespread among the population.

By all accounts, this form of tool use does not appear to bring any advantages. Whether dolphins are 'spongers' or 'non-spongers' seems to make no difference where fitness, life expectancy or successful reproduction is concerned.

All in all, the researchers have found themselves faced with a conundrum.

Whatever underlies the behavioural quirks of these Western Australian dolphins, theirs must be regarded as a material culture,[42] since within it young animals learn from their parents and the number of 'spongers' increases. There is no other plausible explanation to hand.

However, the 'spongers' are not the only tool users. Among the dolphins, there are also 'shellers', who pick up conch shells and use them to eat small fish.[43] It is currently unclear whether the shells are really being employed as tools or whether their function as boltholes where small fish live is simply being exploited. In order to be a true tool, the conch shells would have to be actively used to catch fish. It is impossible to tell from a conch on the ocean floor, though, whether it is simply lying there by chance or if a dolphin has laid it there as a trap, and so the question of tool use here still remains open.

Animal architects

Frankly, drawing is not my strong suit. It is a mystery to me how anyone can portray a landscape or a receding row of houses on a flat piece of paper. How on earth can a person get the three-dimensional world down on a flat, two-dimensional sheet of paper? It's all a matter of perspective, you'll say, and of course you'd be right. Just imagine that you had to draw a street that vanishes somewhere in the distance. The street would be made up of large flagstones, and for you as an experienced draughtsperson, it would not pose the slightest problem to show the street growing ever narrower and the flagstones ever smaller as they recede into the distance. The effect is simple but effective: the paving stones right in front of you are large and the street wide, whereas the stones that are further away are much smaller and get lost in the depth of space. Even a tree shown far off down the street would appear correspondingly small.

Now imagine yourself reversing the size of the flagstones. The street would still recede to a distant vanishing point, but right in front of you would be a mass of small stones, while the larger ones would be further away from you. The optical impression would alter from one moment to the next, and distant objects such as our tree would appear to be somewhat larger. Well, this is exactly the effect that bowerbirds use to make themselves appear larger.[44] Yet the inventiveness of these little birds goes far beyond this, for at the end of the 'street' there stands not just a solitary bird – no, the bowerbird presents itself outside a magnificent edifice.

Bowerbirds are close relatives of birds of paradise, but in stark contrast to them, very few species of bowerbird boast colourful plumage. The great majority are remarkably unspectacular, and their drab feathery raiment and modest size put you more in mind of some ascetic intellectual or a common or garden blackbird. But as their name suggests, bowerbirds build bowers – though, compared to the constructions other animals erect, perhaps the term 'palace' might be more appropriate here, for these birds are genuine master builders. Their constructions have as little in common with other birds' nests as a chocolate biscuit has with a prawn cocktail. The bower's appearance and purpose is quite different: this is a love nest, which like some colourful feather or a red open-top sports car represents the fitness and attractiveness of the male of the species. Yet not just the nest itself but also the way it is decorated plays a key role, and in order to assemble their bowers, the male birds are gripped by an almost manic urge to collect objects. Thousands of items are gathered and, neatly segregated according to colour, laid out in front of the bower; some species even paint the walls with the juice of berries.[45] Some observers claim to have seen the birds using small bundles of dried bark as paint brushes.[46] The finished arrangements are utterly useless, but stunningly beautiful. They fulfil just one purpose, namely to convince the female of the creativity of the builder.

One might be tempted to assign purely aesthetic value to the bower as well, but some researchers who have closely observed the procedure of bowerbird mating suspect that it has another function entirely. In their view, the bowers prevent a potential rape. Initially, this comes as something of a surprise, since one would expect the bower to be rather some kind of a trap, from which the female could no longer escape. But in fact the opposite is the case. After a female, fascinated by the decoration and size of the bower, enters the structure, she is harassed from behind by her suitor. The female now has a surprising advantage, however, as she only needs to turn her head towards the exit. In his eagerness to somehow get her rump between his wings, the male must relinquish his position guarding the entrance, thus giving our unwilling female a free escape route to the outside world. This observation runs counter to the all too frequent instances where rape is a successful reproductive strategy. The researchers explain the evolutionary origin of this unusual anti-rape mechanism by the fact that those males who have constructed especially rape-proof bowers are the ones who are ultimately preferred by the females. In this way, the genes that are responsible for the ability to build rape-proof bowers assert themselves. Some bowers are even more secure, taking the shape of narrow passageways that are almost entirely filled up by a female. If a male wishes to approach, it needs to first walk round the outside of its own construction. This gives an unreceptive female plenty of time to make herself scarce.

Yet these unprepossessing-looking birds have even more surprises to offer. Although the industrious males are given to helping themselves to their neighbours' building materials, in the process quite coincidentally destroying their constructions, they do also pass their knowledge on to juvenile birds. Of course, this training ends abruptly whenever a potential mate appears, but up to that point they are diligent teachers, imparting the basics of structural engineering and the tricks for achieving the optical illusion outlined above, as well as colour theory and the skills needed to design the bower. Incidentally, to human observers, this scene would not look like an organized lesson, but much more like sheer chaos. A group of juvenile birds arrives and tears downs part of the bower, while an agitated teacher rapidly repairs everything. Our own young have mastered this same trick too – haven't you ever asked yourself why children love smashing up things? It's so they can watch and see how it is reconstructed and repaired. In return, they offer us every conceivable help. The same process goes on among bowerbirds, with the young being industrious and selfless gatherers who help the master builder amass a huge amount of material. Once the young birds have mastered the basics, they set about putting together test bowers on their own. These structures do not appear to inadvertently attract female visitors; instead, their sole purpose is for practice and to enhance the bird's knowledge.

An interesting tale, well known among ornithologists, is that of Billy. Billy was a spotted bowerbird (*Chlamydera maculata*) and in common with others of his species he liked to decorate his structures in the colours white, green and red. One day by chance he ended up in a valley where only satin bowerbirds (*Ptilonorhynchus violaceus*) lived, a species that decorates its bowers exclusively with blue objects. It did not take long for Billy to start copying this behaviour.[47] This was quite remarkable, as animals are generally incapable of deviating from a behaviour once it is learned, so demonstrating that a process of lifelong social learning actually occurs in the animal world.

A further indicator of cultural information transfer can be found in the calls of bowerbirds. Presumably it is sexy to incorporate new calls into the vocabulary of seduction, and as a result these little birds, or at least certain individuals that live in a particular area, have developed the ability to sound like a herd of whinnying horses galloping by. But if no horses happen to live in the region, this observation raises a number of questions. One study found that horses had most definitely been present there some fifty years previously; it appears therefore that the bowerbird calls mimicking horses must have been passed down from one generation to the next. As we have learned already when considering the topic of material culture, divergent behaviour in different populations that otherwise live under the same environmental conditions is a strong indicator that cultural information transfer has taken place. In the absence of any other possible explanation, scientists therefore regard a culture among bowerbirds as highly probable.[48]

Bowerbird structures are so attractive it can sometimes happen that a female of another bowerbird species comes to the bower and finds herself drawn into a tryst with the occupant. The resulting hybrids, which are themselves capable of reproducing, have naturally pulled a fast one on evolution, for in truth this mechanism ought logically to culminate in a merging of the various species, yet this does not actually occur. Presumably these birds still hold a few surprises in store for us.

In conclusion, it is worth mentioning that the romantically con-trived love story ends just like it does in real life. After successful mating, our Lothario chucks the female out. The man has more important matters to attend to, for his magnificent edifice would have been a complete waste if it failed to beguile dozens of females. To the credit of male birds in general, however, it should be noted that, in stark contrast to the majority of mammal species, most who sire young are devoted fathers.

Tools without culture

Please don't go thinking that all animals that use tools also have a culture. Nonetheless, the use of tools is a particular achievement, and even 100 years ago nobody would have thought any animal capable of such a thing. Nowadays, it is widely known that corvids, for example, fashion and use tools and that sea otters use rocks to crack open mussels. But did you know that wrasses do that too?[49] They swim around for a while until they have found the ideal stone, which they then pick up in their mouths, carry back and proceed to hurl with considerable force at their prey.[50]

An aphid wasp of the genus *Ammophila* has devised a quite different way of using stones. Like the vibratory rammers that we use to tamp down the ground beneath patios, paths and roads, they flatten the earth around their burrows using stones that are the size of their heads.[51]

Now, you all know Hansel and Gretel from the collection of fairy tales of the Brothers Grimm. In this gruesome tale, two children are abandoned in the woods. But being clever, they mark the path they have taken with crumbs from their last slice of bread.

In the same way, various species of ant such as *Aphaenogaster rudis* also lay trails. The ants behind can then follow the lure to the source of food that they are after. Scientists call this behaviour 'debris dropping', and describe it unequivocally as a form of tool use.[52]

Do you remember our orcas who fished using bait? One animal had learned the trick and the rest had then copied it. Yet the basic idea here was not invented by ants, but most likely by dinosaurs some 200 million years ago. Dinosaurs are the common ancestors of both birds and reptiles, and fishing with bait has been observed in both of these groups. For instance, some species of gull[53] and heron[54] and possibly also certain crows[55] lure fish with berries, feathers or pieces of bread, while the indigenous American species the burrowing owl uses cow dung to entrap beetles.[56] Meanwhile, some species of crocodile[57] lure birds with sticks, which they balance on their noses while submerged. It's then curtains for any careless bird that might choose this stick as a suitable addition to its nest.

Whether all these examples are instances of a culture is, however, a moot point. Were the methods newly devised and passed on to others through learning, or is the technique evolutionary – in other words, did it arise through natural selection? In part V, 'On Thought' (see pp. 169–268), we will hear a lot more about strategic thinking and how good certain birds are at performing fiddly tasks. It may be that Oren Hasson, an Israeli scientist and amateur nature photographer, just happened to be passing by when the aforementioned crow discovered how to fish using a lure. Who knows, in a few years maybe all the crows in the city of Ramat Gan near Tel Aviv will be fishing with bait. In which case, we would have conclusive proof of a culture, as we have with the crows in Japan who crack nuts with the aid of automobiles.

Even when we think we have gained a good understanding of our environment, where animal behaviour is concerned we are still groping around in the dark. Perhaps you'll discover in your own garden or in the pond around the corner or on your next holiday another example of incredible animal behaviour. All we have to do is keep our eyes open.

*

12. The secret language of animals

The language of animals is a mystery to us. I can still recall clearly how, as a kid, I found the idea of being able to speak animals' language really exciting – my hero figure was a Russian who for seven years did nothing but read books and afterwards was able to talk to animals. I was still rather a reluctant reader at that time, and this really spurred me on to read. But can we really learn the language of animals, or are all our best efforts in vain because they do not actually have one? Come with me now into the world of animal communication, and let's find out which species of animal might possess a language, what cognitive preconditions need to be fulfilled, and why I will most definitely never have a conversation with my dog.

As you must surely have noticed by now, I'm fond of using animal videos posted on YouTube to better illustrate what I'm talking about. In most cases you don't actually need to look at the video in question. But where the following one is concerned, I do urge you in the strongest terms to watch the video first before reading on.[58] You'll see an orangutan from a wildlife reserve talking by means of sign language with a deaf and dumb girl on Skype.

So, did you watch it? I have to tell you the video was a fake, but a really good one, as conversations like this have already taken place for real between humans and great ape species. Also, the spoof video was in a good cause, because it was instructive and an incredibly inspired publicity stunt for the environmental protection organization Rainforest Action Network.

For us humans, communication and language are one and the same thing, but when we deal with exchanges of information in the animal kingdom, we need to be more precise. Even protozoa communicate with one another, and ultimately communication is responsible for the fact that multicellular organisms like us exist at all. But protozoa cannot speak. Thus far, we know of only one species that can talk and that is us. Even so, animals can communicate superbly among themselves and also with us, and there are even some species that at least theoretically have a language or are capable of developing one. This is what I intend to concentrate on here.

In all probability, human language is the most important cultural asset of our species, and we can be justifiably proud of our ability to speak. But what does one actually require in order to develop a language? To answer this question, we need first to be clear about two important aspects. You will then be in a position to comprehend why the talk-show favourite Border collie bitch Betsy with her 340-word vocabulary – famous for being one of the world's most intelligent dogs – will never be able to master a language, whereas bats would be capable of doing so.

Audible language

The two most important ingredients of a language are the ability to learn vocally and to understand the pointing gesture. The Border collie Betsy is undoubtedly a particularly intelligent dog, and both she and her human entourage have my deepest respect. To achieve what they do requires a great deal of application, time, patience and trust. Nonetheless, she could never be taught to say 'miaow'. No dog could manage that. Likewise, no cat could ever bark or pig say 'moo'. In their acoustic communication, most animals are obliged to make the sounds that Mother Nature gave them at birth. In other words, they are genetically conditioned and are not in a position to learn even a single additional vocalization. Fortunately, animals also have different pronunciation, and so some species are at least able to determine from a given call what kind of animal made it. Yet most animals cannot even do that. Similarly, most animals lack the ability to comprehend nuanced linguistic elements or to produce them themselves. Despite this deficiency, however, the majority of animals can still communicate extremely effectively.

You may perhaps have seen ground squirrels in a nature documentary. These animals can actually recognize individual animals from their voices. The question is, though, why do they need to do that? Ground squirrels practise division of labour. One of the most important and responsible tasks for a ground squirrel is keeping watch. To do this, they rear up on their hind legs and turn their heads this way and that. If a raptor appears overhead, they utter a cry that signals 'Down into the burrow!', and if a snake comes into view the cry is: 'Up into the trees!'[59] The squirrel sentinels therefore enjoy a high social status and are not required to take part in foraging for food, as it is their job to keep a lookout around their territory. In return, their needs are catered to by their fellow squirrels. But what would happen if one ground squirrel were to light upon the idea of just pretending to keep watch, in order to be fed by the others? It would have an advantage, since it would thereby acquire a lot of food with minimal effort. Animals that are in an advantageous position over others have a higher chance of reproducing successfully, and as a result this behaviour would be genetically strengthened and become a constituent part of the species' evolutionary development. The only stupid thing about it, though, would be that the group was no longer being guarded, and so the individual advantage would endanger the continued existence of the entire group. The social community should therefore not rely upon individuals always doing the right thing.

What mechanisms might prevent such a development? Our little dissembler is thoroughly plausible, for he too utters alarm calls every now and then and in so doing gives every impression of being a conscientious sentinel. The others members of the community would therefore find it really difficult to recognize a liar in their midst. The only way they can assure their survival, however, is by identifying the liar. It's not just important to properly understand the call that is being given, it's also crucial to be able to identify the caller. This ability, together with a good memory, ensures that our ground squirrels ignore sentries who are responsible for frequent false alarms and do not provide them with food. It is therefore to social communities' advantage to be able to identify individual animals and to remember their behaviour. A small achievement for a squirrel, but a great step forward in the course of evolution.

Dogs too are capable of identifying other dogs from their barking.[60] Social life within a wolf pack is evidently so complex that this capacity has evolved there. Dogs can distinguish different human voices as well, of course – an aspect of their behaviour that makes social life with them so pleasant and natural.

And yet dogs are just as incapable as ground squirrels of incorporating new elements into their acoustic repertoire. Most animals vocalize in the usual manner of their species without having to learn their calls; they are genetically inherited. We humans can, for instance, cry right from birth, and even deaf children begin at some stage to produce the sounds of laughter. On the other hand, animals whose calls are not genetically inherited and have to be learned, but which never get to hear those calls because they are deaf or because they have been kept in isolation, will never produce calls themselves. They are lacking not only the auditory input but also the auditory feedback. Some birds – and we humans – need also to hear the sounds that we produce. Only through doing so are we able to reproduce them as we heard them. That's perfectly logical, right? One exception to this, of course, is deaf and dumb people, who struggle and need a great deal of support in their attempts to enunciate human language correctly.

Certain bird species are past masters at learning, and I'm not even thinking of parrots here. Among corvids, there is clear evidence of a cultural transfer of particular calls.[61] It is interesting to note that most calls are passed on in a gender-specific way and hence that a certain proportion of calls are used solely by males or females. Jays, for instance, are known to recognize fourteen different types of call, and have distinct calls to warn of the presence of hawks or owls.[62] These predators each pose a different type of danger and avoiding them demands a particular response. Researchers studying magpies in Sri Lanka have ascertained that they imitate the calls of their natural enemies to warn not only their conspecifics but other animals as well.[63]

However, the basic principle holds good that if I am incapable of learning new calls, then I will also be unable to acquire language. So-called vocal learning is an absolute prerequisite for the development of language, and we may assume with 100 per cent certainty that animals that cannot incorporate any new signals into their repertoire will have no language either. A language is thus clearly a cultural asset, since it is learned from another individual. But are there any animals that could theoretically speak by virtue of possessing the ability of vocal learning?

Indeed there are. In addition to parrots, there are no fewer than three animals that produce noises that sound exactly like us. In the process, one of these animals puts its nose in its mouth, the second doesn't even open its mouth, and the third does the same as we do. The first animal is an Asian elephant by the name of Koshik, which can speak at least four words of Korean.[64] The second animal is called NOC (pronounced 'no-see') and is a beluga whale, which became famous for ordering a Navy diver to get out of the water.[65] Although its voice[66] sounds a bit like children's prattling, producing that sound nevertheless required the whale to drastically lower its natural vocal frequency. The third animal is a harbour seal named Hoover who was well known for greeting zoo visitors with the call: 'Hey you! Get outta there!'[67] All three share a common fate, having been separated prematurely from their mothers and raised by humans. Hoover was found abandoned, while NOC and Koshik were taken from the wild by their future owners. Yet fundamentally it makes no difference whether an animal can mimic human speech; the important thing is its ability to expand its own repertoire over its entire lifespan. Most birds are only capable of learning their songs in their early years. This of course has fateful consequences where language development is concerned, since adult animals are unable to pass on their experience. Among many bird species, it is frequently only the males that are acoustically active. I wouldn't mind betting that there are several people of either gender who would be only too grateful if their partner couldn't speak. But in actual fact in some bird species, as well as in elephants, whales and dolphins, both sexes are capable of lifelong vocal learning.[68]

The absolute professionals in this realm are parrots, first and foremost among them Alex, an African grey parrot named by his owner, Irene Pepperberg, after the 'Avian Language Experiment'. Alex was not only capable of adding words and sounds to his repertoire, but he could also understand their meaning. He was able to count and to use adverbial phrases of manner correctly in their context.[69] He could recognize seven colours, five shapes and could count up to six. When asked, for instance, 'How many rings can you see?', he gave the answer 'four'. If he was then asked: 'Which ring is different from the rest?', he would have been able to respond 'The red one!'[70] Whether he would have developed these same aptitudes in his natural environment remains open. The fact is, however, that grey parrots are extremely social creatures, which, when they are separated from their group, are perfectly capable of forming a close bond with humans. Whether being kept on his own by someone did him any good and whether he enjoyed the company of people is unclear; all we know is that he died after only just half his potential natural lifespan as a result of a heart condition.

Another famous bird of the parrot family, a cockatoo, is Snowball, who is a dancing rock star. His YouTube video[71] has over 6 million hits. Most people find this video amusing, but very few stop to ask themselves why there aren't more animals that can dance. Vocal learning and dancing are made possible by the interaction of two different areas of the brain. Just as with the production of sounds, so must the production of movement be harmonized with what one hears. Accordingly, it is hardly surprising that the only animal species that can dance are the ones that are also capable of vocal learning.[72] Dancing bears do not belong in this category, though, as their dance-like motions were caused by making them stand on a hot plate that burned the sensitive soles of their feet.

Thanks to the anatomy of their larynxes, our nearest relatives the anthropoid apes are unable to articulate language in the way that we do.[73] Our larynx has migrated further down our throat, thereby leaving more room in our mouth for our tongue. In contrast to apes or dogs, our tongue can move in a three-dimensional way. The cavity that this created, combined with our vocal cords, meant that we were anatomically equipped to speak. Our nearest relatives, the great apes, were not so fortunate and are, like most other animals, confined to using the sounds that were given to them genetically.

Bottlenose dolphins – the species of dolphin familiar to us from dolphinariums and the children's TV series *Flipper* – are masters in lifelong vocal learning. Nonetheless, researchers like John Lilly[74] have tried in vain for years to entice them to utter even a single human word. Conversely, it would also be just as fruitless to attempt to teach humans how to whistle like dolphins. No human being can whistle anything like as rapidly as a dolphin. For this reason, the behavioural biologist Denise Herzing uses an underwater keyboard that can synthesize the sounds of dolphin whistles.[75] The instrument includes a small computer that operates with music recognition software similar to the Shazam app.[76] The software is programmed to recognize dolphin whistles and to assign these to objects, in so far as they could be ascertained. In her blog, Herzing wrote that she could scarcely believe it when the instrument whispered 'sargassum'.[77] Sargassum is a brown macroalgae (seaweed) that is widespread off the Bahamas and is popular with dolphins for playing with. It has even given its name to an area of the North Atlantic – the Sargasso Sea. The seaweed wafts about there often in such density that it forms underwater forests. It is said that boats can even get caught up in its tangled mass, and several people consider it a natural explanation for the legend of the 'Bermuda Triangle', where vessels mysteriously vanish without trace.

Imagine someone jumping from their boat into this environment and attempting to communicate with wild dolphins; it would be like a person going into a wood and trying to talk to animals by means of a technical apparatus. Totally absurd, of course, and if the incident made the news at all, it would be a report in some local rag about the crazy eccentric. Not so where Denise Herzing is concerned, however; she is the author of countless scientific articles and an acknowledged expert in the field of dolphin research.[78] Also, her project has some really sound bases, since she has already succeeded in demonstrating that dolphins can understand a form of language. The language in question here was a customized sign language that was developed on Hawaii. The researcher Louis Herman was able to show in his experiments that dolphins can not only use and abstract new terms without difficulty, but that they are even capable of comprehending short sentences with predetermined grammar and behaving accordingly.[79] The sign language that he personally devised became increasingly abstract over the course of the experiments. To begin with, the dolphins saw humans on the side of the pool using the sign language. Later, all they could see was a monitor behind an underwater window, and finally the sign indicated was reduced to two white gloves on the monitor. In addition, the trainers wore black clothing against a black background so all that could be made out on the videos were the moving hands.[80] This was unquestionably a high degree of abstraction.

But how could one test whether this wasn't in fact just a question of a particularly successful training regime? Furthermore, it was necessary to prove that the terms learned were really being understood for their meaning. To find this out, Louis Herman employed a method that we are all fond of using to test if a person is truly paying attention and has understood us. The following is a simple example: the dolphins were able to carry a particular object to a particular place. In other words, one could say to them: Take the red ball to the square basket. Yet they also knew terms for other objects from their surroundings. Thus it would have been possible to issue the following command: Take the round basket to the blue ball. Or: Take the blue ball to the underwater loudspeaker. But what would happen if they were ordered to take the underwater loudspeaker to the square basket? Animals that had just been trained to carry out orders would at this point try to manipulate the underwater speaker. But the dolphins in the study reacted differently. They were fully aware that the underwater loudspeaker could not be dismantled, and so they did not swim towards it, but instead looked inquiringly at their trainer in the hope of perhaps receiving a more sensible command.

The experiments in understanding the concept of speech went even further.[81] An especially gifted female dolphin called Akeakamai was asked questions, which she answered by pressing one of two buttons marked Yes and No. After she had grasped the concept of Yes and No, she was also capable of answering the following question: Is the blue ball in the pool? If it wasn't there, she responded with

No. Most animals cannot do that, since they have no conception of something that is not present (see 'Mental images'; see pp. 171–5). The vast majority of animals can follow a trail with the utmost perseverance, say, or search for a particular familiar object, but they cannot articulate that the object simply isn't there. What's more, Akeakamai's achievements are not an exception among dolphins; other researchers have made similar observations.[82]

The aforementioned wild dolphin researcher Denise Herzing was therefore attempting to confirm the hypothesis that the linguistic abilities of dolphins observed in captivity were also deployed in the open ocean. Perhaps you are asking yourself why one cannot simply observe this from interactions among the dolphins themselves. Indeed, this would be the ultimate dream of many researchers, but there is one slight problem. Like our beluga whale when mimicking human speech, dolphins do not have to open their mouths in order to produce calls. If I were to shoot an underwater video and spotted a group of dolphins, it would be impossible to discover which dolphin was saying what.[83] Another factor is that dolphins can produce a variety of sounds. The most intensively researched of these are the whistles, which we'll discuss further in the chapter 'Who am I, and who exactly are you?' (see pp. 196–230). Dolphins also make clicks, which are perceptible only in the ultrasonic region that is inaudible to us, and a range of squeaking sounds. Another problem is a general one of bioacoustics: when the recordings are analysed, it is unclear which sounds are just background acoustics and which are the actual signals. The recordings were evaluated both by humans and computers and what at first sight appeared straightforward turned out to be incredibly complex.

Interestingly, one fruitful approach has been to try to solve the problem in collaboration with astronomers who are engaged in the search for extraterrestrial life. Researchers working in this discipline experience the same difficulty: they find it hard to distinguish between white noise and signals.[84]

Finally, we should not forget that a group of dolphins is practically constantly in motion within a three-dimensional space. This is also a challenge for scientific research into parrots,[85] meaning that the two groups of animals with the highest probability of having developed a language cannot be easily studied in the wild. It is far simpler to study a group of great apes sitting relatively still at one particular location than a group of dolphins or birds. As a result, we will surely have to wait a long time before we can converse with them.

Yet there are some bird species that are far less difficult to research in their natural surroundings. Every evening I sing a lullaby that goes: 'Little tomtit, little tomtit, tell me where do you come from?' to my small children to send them off to sleep. The song is about hungry tits in winter and I sing it with the uneasy feeling in the back of my mind that I'm actually opposed to feeding wild animals. Too often, feeding has a negative impact on the environment or on behaviour. You can see this easily by visiting any town pond. Among ducks, human feeding increases aggression to a wholly unnatural degree and in summer many bodies of water begin to smell because not all the morsels of bread get eaten, and sink to the bottom where they rot. In the case of our native songbirds that have chosen to overwinter with us, it's absolutely permissible and desirable to feed them during spells of harsh weather, so in fact I can square my conscience with the lullaby after all. However, after coming across an article of enormous significance in the course of my research, I now sing the song with other thoughts in my mind. Despite its importance, the article was barely picked up by the press, and nor did anyone in my circle of friends and acquaintances alert me to it. For the very first time in an experiment conducted in the wild, researchers in a mixed forest near the city of Karuizawa in Japan had actually gathered evidence to prove that animals are capable of using syntax – that is, sentence construction in accordance with a set of grammatical rules.

And the subjects of this groundbreaking study were none other than our familiar coal tits, which spread from the Far East to both Europe and the rest of Asia. These lively little birds were played recordings of their own songs. The songs contained different injunctions. Thus, the sequence of sounds A B C meant 'Look out!' – a kind of alarm call intended to induce greater watchfulness. Then came a sequence of sounds designated D, which meant 'Come here!' The birds also liked to combine A B C with D, which denoted: Take care and come here. If the sequence A B C – D was played to the tits, they behaved in accordance with the injunctions. By contrast, if they were played D – A B C instead, nothing happened.[86] This 'sentence' simply made no sense to these small birds, which each weigh around 15 grams (½ ounce).

This was an incredibly simple experiment with incredibly far reaching implications. We are talking here about the first proof obtained from the wild for the rudiments of a genuine language among animals, and bear in mind we are 'only' dealing with tits. But why am I at all surprised? After all, we learned earlier that animals *can* inhabit a culture. Such moments make me realize how wrong our whole conception of the lives, thoughts and actions of animals has been to date. In any event, we are slowly beginning to comprehend how a language could evolve in the first place and why animals have begun to expand their portfolio of signals, and how vocal learning came about. You'll no doubt be surprised to learn that even our tiny house mice are well on the way to developing a language. As we are going to discover, everything begins with dialect.

*

Dialects in the animal kingdom

We humans find it relatively difficult to conceive of something as half-complete or in gradations. For example, I cannot think of anyone who could explain to me in even a halfway lucid manner what a state of semi-self-awareness would feel like. In these sorts of matters, we only recognize yes and no, and black and white. It's a very similar situation with vocal learning. We imagine that an animal can either learn new signals or it can't. We like being able to draw a clear line between things – in this case, between animals that are definitely not capable of acquiring a language and those that might at least theoretically be able to. Nature is far more complex than our powers of imagination, however; something that is inconceivable to us can be perfectly normal in nature. Dialects are a prime example of this.

When we talk about a dialect, we are referring to small nuances of our language that mark us out as a member of a particular group. For instance, we live in a particular area where, say, Geordie (Newcastle English) or Scouse (Liverpool English) is spoken. It's a similar story with slang, which is typically spoken by a particular social grouping. This shading of our language engenders a sense of commonality and allows people to demarcate themselves from others. As part of a community, one feels oneself thereby strong and secure. In all likelihood, it was precisely this need for security, strength and protection that first led to the sounds that people were born making being altered ever so slightly. This made it easy for people to recognize one another and the key technology of vocal learning opened a window to the acquisition of language.

Orcas: Soon after I started studying animal behaviour, a publication about dialects in orcas caused a massive furore. Up to that point, the idea had been regarded as pure nonsense, and I can vividly recall the criticism that was levelled at the first papers to appear on this topic. To gain a better understanding of this discovery, we must examine the social life of orcas. The orca group that has been most exhaustively researched, thanks to their attachment to a single location, are the 'residents' off British Columbia on the west coast of Canada; these animals live in small matrilineal family groups of up to eight individuals. Each of these family groups is led by an experienced mother animal. And up to three such families form a 'pod'. Multiple pods make up a clan, and each clan has its own dialect.[87] In principle, this is rather like a small, close-knit village community whose nuances of pronunciation distinguish it from a neighbouring village. Certain calls remain unaltered and stable over several generations, whereas others undergo a change.[88] After the dialects had been discovered off the coast of Canada, the calls of orcas living along the Russian Pacific seaboard were also studied. In the course of this investigation, it turned out that the animals not only knew dialects but also differed in the basic characteristics of the sounds they made.[89] The researchers compared these sounds with our phonemes. So, for instance, English does not have the voiced 'kn' sound that is common in German, while the Japanese cannot pronounce 'st'. The ability to recognize such differences is possessed not only by humans, it is also demonstrated by indigenous seals when they are confronted with the sounds produced by different orca groups. Logically enough, fish-eating orcas pose no danger whatsoever to them, whereas they react with panic when they hear the sounds of passing mammal-eating orcas.[90]

Thus far, however, we have no idea what the actual purpose of these dialects among orcas might be. Since only individuals with different dialects mate with one another, one could speculate that their role is to avoid inbreeding. On the other hand, I have no doubt that the animals would recognize their closest family members without the aid of dialect. Likewise, the hypothesis that the orcas can use dialect to announce their location across great distances is not wholly convincing either, for in all probability they can also recognize one another from their individual voices. Furthermore, there were some surprising observations that showed the animals imitated the dialects of other groups. They could even mimic the sounds made by other species, as researchers noticed that orcas that had been captured and kept isolated from their conspecifics began imitating the sounds of sea lions[91] and dolphins.[92]

All in all, though, we are faced with a conundrum: we can identify neither the basis of the orcas' culture nor any reason for the evolution of dialects. All we know for sure is that there is no simple explanation for the behaviour observed. By implication, we might even say that their social life is simply too complex for us to grasp.

Birds: It is well known that many bird species are outstanding mimics and complete naturals in the art of vocal learning. It might therefore come as no surprise to learn that we also encounter dialects within this species-rich group. Especially among migratory birds, we also have a thoroughly convincing explanation for the formation of dialects. It is, for instance, conceivable that after migrating to the regions where they customarily spend winter or summer, the birds might recognize one another again by their dialects and then band together to exploit and defend resources. Researchers have ascertained that starlings share their roosts with starlings with the same dialect,[93] and corvids were found to adopt the vocalizations of others. A captive pair of ravens was used to study this phenomenon. Wild conspecifics were then exposed to this pair of birds, with the scientists observing in what form elements in the vocalizations of the captive pair were absorbed into the repertoire of the wild birds.[94]

Bats: Though many people might think of bats as some kind of flying rodent, they actually have more in common with whales and predatory mammals. However, their phylogenetic history is not the point at issue. Rather, it is their social life that is of interest here; often, different bat species live together peacefully. This is extremely rare behaviour in the wider animal kingdom, and indeed I can only think of one other example. Dolphins and whales often swim in groups with other species. Transposed to the human world, this would be tantamount to us living alongside or undertaking migrations with other anthropoid apes, namely gorillas, orangutans and chimpanzees. The groups in which bats live are often huge, numbering several thousand individuals. Bracken Cave near San Antonio in Texas, the largest bat colony in the world, is home to around 20 million animals. That's as many bats as there are people in New York City. Within this teeming mass, it is of vital importance to have access to smaller, more manageable networks. Possibly for this very reason, the animals have developed a variety of different dialects. One very recent finding is that they are even capable of adopting the dialect of another group.[95] Lest anyone should still be in doubt over whether animals are capable of vocal learning, this should dispel such notions.

Another enormously exciting aspect is the fact that, like the ground squirrels we discussed on page 77, the bats can recognize one another from their voices.[96] I'm hoping, though, that no scientist ever manages to prove that all the 20 million bats in Texas know and recognize one another. For if that were the case, we humans would be toppled from our top spot in terms of intra-species recognition. Most people know just a few hundred fellow humans, but depending on lifestyle and occupation, that number can of course swiftly grow. My email address book contains nearly 3,000 people, but if it weren't for my notes against individuals' names I would have no idea who many of them are.

Anthropoid apes: As we have already noted, the three species of great ape other than ourselves are not especially good at vocal learning, since their speech apparatus simply does not allow it. Researchers were therefore all the more astonished when they analysed various groups of chimpanzees and discovered calls that were specific to particular groups.[97] Because the variety in the calls was not genetically conditioned, only one possible explanation remained, namely that the animals had learned the calls and were therefore capable of vocal learning, albeit only to a limited extent.

Elephants: As the example of our talking elephant Koshik showed, elephants are also capable of vocal learning. It is no surprise that the animals also have a good sense of rhythm. In Lampang in north-eastern Thailand, there is even an Elephant Orchestra.[98] Unlike the phenomenon of elephants using their trunks to paint with a brush, which is a pure sideshow as elephants are colour-blind, such an orchestra actually makes sense. Even though I can say nothing about the conditions in which the animals in Lampang are kept, elephant music is a good example of their complex cognitive abilities.

House mice: There's no doubt that the group of vocal learners is a very small, almost select band of cognitively highly evolved animals. Yet would you have ever thought it possible that our humble house mice might be among them?[99] Among them too, appropriate neural feedback mechanisms ensure that they can hear the sounds that they produce themselves. In this way, they are able to replicate new calls exactly as they have heard and learned them. They use this capability to generate group-specific dialects, while the males employ it to promote their qualities as suitors by 'singing' to impress the females.

But it's not just the males that call. In 2016, a small working group at the Max Planck Institute for Evolutionary Biology in Plön in northern Germany published an article entitled 'Communication at the Garden Fence'.[100] They did not mean by this some inconsequential chit-chat across an allotment boundary, but instead a context-dependent conversation employing various different 'words' and conducted as part of social interaction between female mice. Yet how was it that this finding remained hidden for so long? As is so often the case, we simply weren't looking or listening properly. Communication between mice occurs largely at frequencies that are inaudible to us. Elephants and baleen whales can communicate within the infrasound range – in other words, at lower frequencies than we can hear – while bats and dolphins use the ultrasound range, that is way above our hearing capacity. Mice belong to the latter group, and their calls are simply too high-pitched for us to hear. So-called ultrasound vocalization (USV) was simply ignored. The researchers from the Max Planck Institute, who kept two large mouse groups within a compound measuring some 20 square metres (215 square feet), so providing them with something approximating to a natural living environment, ended their report by concluding that communication among mice had been rather underestimated hitherto. In the end, this dispassionate observation means that we still have a lot of surprises in store where mouse language is concerned.

At present, researchers are trying to gain a better understanding of the social network in which the animals live.[101] (For more on this, see the chapter 'Like Facebook, only different'; see pp. 137–47).

Humans: There's no question that we humans are the champions at vocal learning. Yet until recently, it was assumed that the ability to imitate began in the cradle and was therefore innate. However, this is probably not so, for it appears that we only learn this skill in the first weeks and months of life.[102]

Talking dirty

Exclamations like 'Ooooh!', 'Aaaah!' and 'Yes!' might be turn-ons in certain intimate contexts, but they don't really fit the bill as dirty talk. They are noises, however ardently and convincingly they are uttered. Instead, something like 'Hey there, you!' whispered in lascivious, husky tones, or 'Hold me!' – accompanied by a provocative look – are nearer the mark, and can be used as a springboard for more raunchy stuff. So, when we refer to 'talking dirty', we really do mean speech, and anyone who speaks in whatever context avails themselves of a common currency of clear signifiers.

For a moment, just picture yourself in a strange land where you don't speak the language. You're standing in the middle of an exotic market, and would dearly love to taste one of those yellowish, roughly plum-sized fruits that you can see hanging in bunches from a wooden shelf. Naturally, you are an inquisitive and forthright person, and you don't shy away from signalling your wishes by pointing at the fruit. In an extended gesture, you also jab your finger towards your open mouth. Of course, the vendor gets your meaning and hands you a longan. You crack open the shell and, hmm, it tastes like a lychee but the heart is horribly bitter. In this situation, you have posed a question, been understood and as a result have achieved your goal. Without knowing it, you have used patterns of behaviour tried and tested over millennia, and purely theoretically you could communicate in the same way with our nearest relatives, the great apes.

All chimpanzees, gorillas and orangutans use around a third of their genetically inherited signals and gestures in common,[103] and like us they are all great apes. We humans understand these gestures even though we no longer actively use them.[104] Gestures accompanied by little mime-shows in all probability represented our very earliest stages of language acquisition.

Some researchers are now speculating whether comprehension through body language as a preamble to sexual congress might have been the origin of human language.[105] For instance, female bonobos, a great ape species, were seen to flirt with one another in this way. A pointing gesture and an unmistakable swing of the hips anticipating the sexual act were all it took to get the party started.

Together with the aforementioned leaf-clipping, it is by no means fanciful to suggest that such dirty talking may have represented the origins of our language. Once again, then, this would mean that sex stood right at the centre of our motivation, and we would have good cause to look more closely at body language.

Gestures and symbols

Even though animals cannot speak human language, there is one animal that managed to learn a human language and was able to communicate admirably using it. The animal in question was called Washoe, a female chimpanzee, and the language she learned was ASL (American Sign Language).[106] She was captured in 1965 in West Africa in order to serve as a laboratory animal for the US Air Force. Just two years later, the researcher Beatrix Gardner began tutoring her in ASL.[107] The unusual thing about her training was the way people treated the chimp. She was allowed certain free spaces and her feeding was separated from her instruction. This is unconventional, since normally wild animals only do something when there is a direct reward. You can see this clearly in dolphinariums, because although if properly treated dolphins are capable of performing certain 'tricks' voluntarily, generally speaking in the entertainment shows put on in zoos, they are given fish for their performance. If you want wild animals to put on a show, it's just much simpler to feed them during it. So, dispensing with feeding during training was something of a risk and only made possible by a great deal of trust on both sides. This trust paid off, as Washoe proved to be a good pupil, eventually mastering more than 130 signs. Ultimately, she amazed her trainer and researchers studying her in equal measure when she independently signed 'water-bird' to denote a swan, thereby combining two concepts that she was familiar with into a new one. She also gained a certain fame in the media, and so it came as no surprise that the *New York Times* printed an obituary of her.[108] Yet Washoe was no isolated case, with other animals also surprising researchers with their new coinages. Thus, a hot radish was signed as 'cry-hurt-food'[109] and a Fanta as 'Cola that is orange'.[110]

The primatologist and psychologist Sue Savage-Rumbaugh took a different approach, devising her own language that she called 'Yerkish', which was based on symbols.[111] In this, the symbols (lexigrams) are not icons but instead are fully abstract; in other words, it is not possible to divine the meaning from the picture that is displayed. A bonobo male by the name of Kanzi is perhaps the most famous speaker – or more accurately user – of this language. He can use almost 400 lexigrams.[112] His behaviour has been written about in nearly 5,000 scientific papers.

In summary, it is clear that both dolphins and great apes are capable of understanding and using simple three-word sentences. However, it remains unclear whether they would ever avail themselves of these capabilities in the wild. And yet some surprising proof did come to light of a syntax being used among gibbons (the so-called lesser apes).[113] Apart from us humans, they are the only primates that can boast an abundance of acoustic signals. Their calls are frequently described as songs.[114]

★

Body language and pointing

As well as acoustic communication, information is exchanged in the animal kingdom in every conceivable way. Body language is one of the most familiar of these ways: on the one hand because we use it ourselves, and on the other because it is easily perceptible to our senses. Among our nearest relatives the great apes, almost a hundred gestures have been identified, with a great diversity of meaning, and body language plays an important role among many other animals as well. At this point, though, I'd like to concentrate on a single gesture, one that forms the most fundamental component of each and every language. This is the pointing gesture, which comes so naturally to us humans. In recent years, this gesture has sparked a far-reaching scientific dispute.

Not long after completing my studies, I spent several weeks abroad with my wife. We didn't want to subject our dog, Captain Flint, to a series of stressful flights, so we decided to leave him in the care of some good friends of ours in Kiel. On our return, our friends naturally gave us an exhaustive account of Captain Flint's time with them and all the things they'd got up to. We were standing in the kitchen, and Flint was staring fixedly at a corner of the room where a basket containing all the kitchen veg was hanging. Of course, I immediately twigged what my dog so desperately wanted and so I asked the lady of the house whether I could feed him a carrot. She looked at me wide-eyed for a moment and then burst out laughing. In response to my puzzled expression, she replied, while still gasping for breath: 'That's incredible! Every day, Flint went and stood in the kitchen and stared up at the wall but it would never have occurred to me he was after a carrot.' I didn't feel at all like laughing at that point, because I realized I'd clean forgotten to tell her before we went away that every night, just before we went to sleep, we'd let our 75-kilo (165-pound) canine companion come on to our bed and crunch on a carrot before retiring happily to his basket.

My wife and I didn't realize at the time what an amazing ability Flint had displayed at our friends' house. Pointing – that is, staring at something or using a hand to indicate an object – is the first step towards abstraction. Imagine yourself taking a stroll around town with some friends. Unfortunately, they walk past your favourite ice-cream parlour because you've all just eaten dinner together. You didn't much like your meal, however, so you left the restaurant still hungry. The weather's warm, and the ice-cream parlour promises to totally redeem your bad experience, and you can almost taste the blissful explosion of flavour that awaits you. The easiest way to achieve your aim is to raise your arm now and point at the ice-cream parlour. Until very recently, animals were thought incapable of doing such a thing.[115] The fact that well-trained dogs can also understand pointing gestures made by humans was attributed to their ability to adapt to living with people,[116] since no equivalent capacity could be observed in wolves.

Some while ago, I was asked to write a chapter on communication among dolphins. In the book where it appeared, published by the University of Freiburg, there was also a chapter on communication between humans and horses, by the researcher Marion Mangelsdorf.[117] My knowledge of horses is of a purely theoretical nature, and so her meticulously detailed description of her interaction with horses left a deep impression on me. And yet I would never have dreamed of ascribing to horses any capacity for pointing; but recently I was surprised by a publication that described horses as behaving exactly like Captain Flint. And that wasn't all – they also attempted to make contact with their owners, and once they had established eye contact, they would glance down quickly at their feed bucket, which was beyond their reach.[118]

You won't be surprised to learn that dolphins are also able to understand[119] and use[120] pointing. Similarly, our nearest relatives the great apes are capable of following gestures[121] and of making their wishes known through gesturing.[122] There are even some suggestions that this might also apply to ravens.[123]

Hitherto the situation was straightforward, but I feel duty bound to reveal to you that an almost vehement debate has erupted in recent years over pointing and the evolution of human language among great apes. Of all the animal candidates that might conceivably possess a language, the anthropoid apes have been the most thoroughly researched. They have been subjected to countless experiments in captivity and also intensively studied in the wild. On the one hand, research on great apes is relatively simple, and on the other their genetic proximity to us humans is especially attractive to scientists interested in how our language evolved. As such, it is hardly surprising that a number of contradictory observations have resulted and the more one knows, the more questions arise. Let us therefore take a closer look at the current debate.

Even the gist of this debate is really difficult to grasp for someone like me who has not been schooled in philosophy. Let's begin by thinking back to our example concerning the ice-cream parlour. You have gained the attention of your group and pointed to the object of your desires. But what precisely has taken place within you, and between you and the members of your group? Fundamentally, there are two possibilities: although you have sought the attention of your group and pointed to the ice-cream parlour, you have remained within your own world. You felt the need to remain within your group and therefore drew attention to yourself. Beyond this, you also felt the need for ice cream and so pointed to the ice-cream parlour. However, it is uncertain whether the others recognized your desire to eat an ice cream – after all, they could not know that you were still hungry. In such cases, one speaks of intention.[124] This frequently involves a deliberate or planned or perhaps also conscious action. However, it occurs with no knowledge of whether others recognize or even share your desires. In contrast to this stands so-called ostensive-inferential behaviour. Communication in this mode is based on the principle of cooperation, and is named after the linguist Paul Grice (Gricean communication). It assumes that communication among people equates to cooperation, which all participants abide by in order to agree on a particular aspect. In other words, all those involved know that the subject in question is an ice-cream parlour. You therefore know, for example, that your friends understand that you want to eat an ice cream. Otherwise your friends would only comprehend that you wanted something, but the ice-cream parlour would not be part of their commonly shared reality. Furthermore, a distinction is made between an imperative gesture, namely one expressing a demand, and declarative pointing. In the latter, a particular attitude to an object is communicated. For instance, if we point at a beautiful sunset we are not trying to put across the idea that we would like to eat the sun. Flint the dog, though, did want to eat the carrot and therefore pointed solely in an imperative way.

Views on this are wildly divergent. For some, gestures are only ever intentional, while for others they are ostensive-inferential. Some scientists maintain that we cannot know for sure whether animals agree on a commonly shared reality,[125] while for others this is created when, for example, a cooperative relationship is established through eye contact and the ensuing situation consequently obeys the principle of cooperation.[126] Yet others insist that animals have the faculty of metacognition – that is, the potential to think about their own thoughts (see 'The marshmallow test'; see pp. 192–6), or they apply to them a 'theory of mind', namely the capacity to recognize others as living beings and to empathize with them. Another important point is the question of whether gestures have been learned or inherited. I have already demonstrated that the abstract action of leaf-clipping, that is the crumpling of dried leaves, is used differently in different populations and that as a learned behaviour this equates to a tradition or a cultural legacy. We will deal with other issues such as metacognition or the theory of mind in parts V and VI, entitled 'On Thought' and 'Sentimentality' (see pp. 169–268 and 269–82). After reading those chapters, you may well wish to draw your own conclusions.

Quite apart from all these considerations, the fact remains that animals are extremely good at communicating with one another. Like the ground squirrels we looked at earlier, many animals have calls with different meanings, and among some animals the form of communication is even learned – in other words, it is a cultural legacy, and as we have seen from the example of the great tits, sometimes even a correct syntax is required. But amid all this uncertainty, one thing is beyond all doubt: if we look more closely, we will surely find many other examples of animal cultures than we could possibly even imagine today.

13. What's culture got to do with conservation?

Some ideas are simply better than others, and in the case of certain ideas one even gets the feeling they may be the best thing since sliced bread. That's pretty much what happened to my WDC colleague Philippa Brakes from New Zealand when she hit upon the notion of proposing a scientific workshop on the topic of cultures among whales. The charity WDC (Whale and Dolphin Conservation) has for many years had a very good rapport with the UN Convention on the Conservation of Migratory Species (CMS). This international environmental agreement is focused on animals living in the wild that are no respecters of human boundaries. They often fall through the net of nature conservation measures taken by nation states, since the animals in question are not native species, but instead cross on foot, overfly or swim through the territory of each respective country. These include such diverse animals as the cheetah, various species of sturgeon, albatrosses and saltwater crocodiles. Contact with the secretariat of the Convention in Bonn, reinforced by many collaborative efforts such as the 'International Year of the Dolphin' for example, enabled the WDC to integrate such a workshop into the official working programme of the Convention. Thanks to the potentially high degree of efficacy this kind of workshop promised to deliver, all the big names in the field signed up to be part of it. What resulted was a hitherto unprecedented resolution, which on the one hand stressed the enormous importance of cultural aspects in conservation measures and on the other urged UN member states to take such factors into account.[127] Yet why exactly should cultural factors play any part at all in nature and wildlife protection?

As we have seen, within a cultural community the transmission of information and knowledge is essential and can even be vital for survival. Consider elephants, for example. Among African elephants it emerged that the groups that were led by older females were more adaptable and that their group behaviour was based on a greater wealth of experience.[128] As you can well imagine, poachers prefer large tusks, since these yield larger, more imposing and hence more valuable ivory carvings. Unlike the Indian elephant, even the cows among African elephants sport large tusks. One tragic consequence of this is the selective killing of the older, larger and more experienced animals. Researchers now believe that throughout the world there is not a single elephant that in its childhood has not been forced to witness a close family member being slaughtered. The scientists also think that poaching has destroyed the great diversity of elephant cultures and that most surviving animals are traumatized and exhibit disturbed behaviour. Anyone could claim that, you might be thinking. Yet these findings were published by the world's leading experts in the journal *Nature*.[129]

Although there are still enough elephants worldwide to maintain a genetically stable population, this population has been robbed of its culture and the foundations of its social behaviour. From the point of view of population management and traditional nature conservation this would not pose a problem, since suitable measures could be taken to increase the population once more. But we humans are just beginning to realize that nature conservation simply at the level of populations is not enough for certain species. We must bear in mind that individual animals can be responsible for the survival of entire populations, as bearers of important information. For this reason, I see the resolution mentioned above as a first step towards a paradigm shift, whose focus would be on taking individual animals and their role within a system into account before engaging in any interventions in nature. No distinction would be made here between killing animals to eat them, capturing them in order to display them in a zoo or, as we shall see in the following example, killing them by mistake in the course of military exercises.

Based on findings in the fields of communication, social behaviour and spatial distribution,[130] the Canadian government decided to treat the group of 'southern resident orcas' off British Columbia as an independent protected resource and placed them under special conservation measures. Now, the southern resident orcas know their own circulation area, but of course know nothing about the border between Canada and the United States and so cheerfully meander all over the place. Unlike their Canadian colleagues, however, the National Marine Fisheries Service – a branch of the US Department of Commerce, and the highest marine protection and fisheries authority in the USA – was not in a position to recognize different groups of whales.[131] For this agency, it was advantageous for all orcas in this maritime area, including the Canadian residents, to be declared one single large population. Through long-established computational models of population management, it then became possible to factor in the death of a number of animals during manoeuvres by the US Navy without having to fear any legal consequences. The Americans took the simple view that there were plenty of whales around. Fortunately, following massive protests, this ruling was revoked in 2005, and the southern resident orcas have now been officially declared a specially protected population in the United States as well as Canada. In essence, the US authorities had originally failed to take into account the cultural aspect in two respects. For one thing, the separate Canadian group was not recognized, and for another it was deemed acceptable for a certain, albeit very small, percentage of animals to be killed in the course of military manoeuvres without considering whether one of these might be an important animal for the wider population.

In modern environmental protection, a clear distinction is therefore made between Evolutionarily Significant Units (ESUs) and Culturally Significant Units (CSUs).[132] Unfortunately, even here the significance of particular individuals is still disregarded, and so I hope the resolution will come into full effect in due course.

However, there is yet another consequence of our actions. As you are no doubt aware, many zoos are making efforts to breed and reintroduce to the wild species that are in danger of extinction. Unfortunately, in the past, this has led to problems with conservation, since these reintroductions only work with animals that do not live in a social community with a culture.[133] The reintroduced animals lack any cultural legacy, and often it proves impossible to reintegrate them into the existing social network that has evolved.

IV

A Sense of Community

GENERALLY SPEAKING, a social life is associated with people or animals. That this isn't necessarily the case is demonstrated vividly in Peter Wohlleben's bestselling book *The Hidden Life of Trees: What They Feel, How They Communicate*. In fact, though, one can even top that: did you know, for instance, that bacteria decide democratically to commit mass suicide? You didn't? Well, then you can count yourself among the 99.9999 per cent of people who have no inkling of these hidden goings-on, and you're surely no microbiologist.

Myxobacteria live in the soil and feed on dead organic matter. Unlike animals, bacteria cannot ingest their food in order to digest it. They have no mouth, nor can they envelop food particles with their bodies like many single-celled organisms do. Instead, they produce a mucus, surrounding themselves with it and pumping their digestive enzymes into it. Broken down into its individual molecules, the food is then absorbed through their surface, in much the same way as happens in our gut. However, the bacterial mucus does not only function as an external 'digestive organ', it also guards against drying out and acts as a communication medium. If food becomes scarce, myxobacteria signal 'hunger' using messenger molecules, thereby setting in train one of the most remarkable processes on our planet.

Bacteria do not have an easy life – in fact, to be precise they live constantly on the very edge of minimum subsistence level, and their general watchword is: divide or die. However, uncontrolled reproduction is vital for the survival of bacteria. Although some bacteria are, thanks to a rotating motor and a propeller propulsion[1] that are unique in nature, the fastest organisms on Earth (in relation to their body size), this speed is of no benefit if they cannot find suitable food. Yet if they do succeed in generating a bacterial lawn and communally producing a layer of mucus, they are capable of breaking down the most complex chemical compounds and changing almost everything into food. But even the most abundant food source dries up one day, and then the great extinction begins. Nature has to counter this with something, otherwise all life would quickly be at an end.

One strategy unique to bacteria has been devised by the myxobacteria. They have a 'democratically organized social life'.[2] Of course, they also obey the dictum: feed–divide, feed–divide and feed–divide. But if food grows scarce, they close ranks and alert one another to the lack of sustenance by means of a messenger substance. If this has accumulated to a certain level in the mucus layer, they trigger a phase of life that one would not normally have thought bacteria capable of. They link up and after about a day form a multicellular organism comprising around 100,000 bacteria cells. That's at least a hundred times as many cells as in the threadworm species *Caenorhabditis elegans*, and that has a mouth, a gut, an anus, sexual organs and a primitive nervous system made up of around 300 cells. The so-called fruiting bodies are around 0.1 millimetre in size and can even be seen with the naked eye. Now, most of the cells die and in so doing supply the cells inside the fruiting body with nutrients. These in turn form enduring spores that are carried away by the wind or passing animals until they again encounter more favourable living conditions.

I have set out this particular example in such detail because I want to make it abundantly clear that the invention of social life was absolutely essential for the evolution of life on Earth. It was the social life of protozoa that first made the emergence of multicellular organisms possible, and the social life of multicellular organisms was in all probability responsible for the development of our complex brains (we will go into this topic in greater detail in the chapter 'The thinking apparatus'; see pp. 241–50). In turn, these complex brains have certain essential requirements in the same way that air is vital for breathing – namely, repetition and training. Like people pumping iron in muscle factories, our nerves need to repeat the same procedure time and again in order to become fixed memories, mental faculties and cognition. The only question is: how do you get someone or something to repeat an action over and over again? That's right – by making it fun, perhaps the most important motivation for animals and humans alike.

14. The hedonistic society

For some years now, on balmy summer evenings a song thrush has made a habit of perching on the eaves of my house and sending its song trilling down the valley in front of the veranda. Is she revelling like me in the view and the sound of her singing, or is she just following an inner programming? To answer this question, we should perhaps consult one of the cleverest creatures on our planet. The animal in question is a kangaroo, and it lives in Berlin. The talking marsupial is the fictional flatmate created by the absurdist stand-up comic Marc-Uwe Kling, and a key aspect of its take on life is that the only relevant criteria are 'funny' and 'not funny'. From the point of view of behavioural biology, Kling has hit the nail on the head here. To reduce it to the simplest of formulae: jokes are funny and we engage willingly in anything that is fun. And if something's enjoyable, then we do it frequently. Effort has its own rewards, and so from a biological perspective this willing engagement helps us achieve our ends, be it a full stomach, sexual gratification, delight in our offspring, our desired position within a hierarchy – or, like my song thrush, just the pleasure of a beautiful view.

From this perspective, the hedonistic society doesn't seem so contemptible after all, and we humans will discover kindred spirits in this regard in many animal species. But for now, I invite you to accompany me to any public convenience you care to name:

A man is sitting in a public toilet. Suddenly he hears a voice from the adjoining cubicle: *'How's things?'*

Man: *'OK, thanks.'*

Voice: *'What are you up to right now?'*

Man (mildly irritated): *'I'm sitting here on the loo, just like you.'*

Voice: *'So what are you doing later today?'*

Man (tetchily): *'I'm planning to relieve myself and then do some more shopping.'*

Voice: *'Can I come over and see you?'*

Man (now furious): *'For God's sake! Lay off or I'll call the police.'*

Voice: *'Look, can I call you back? There's some maniac in the next cubicle keeps jabbering away at me and now he's threatening to call the police as well.'*[3]

There's a fine argument to be had about what constitutes humour, certainly, but over and above this, all jokes have a common denominator. You're regaled with a shorter or longer anecdote and then all of a sudden, you're told something that's supposedly just another element of the story but after a split second you realize there's an incongruity and start laughing. The humour comes from the surprise in the plot twist or an inversion of the basic premise.

In the past I have often asked friends and acquaintances what they think the greatest difference is between human beings and animals. Most responses don't faze me and I always have a convincing counter-example to hand. But one of my interlocutors managed to trump my biologist's logic one time when he replied: '*Animals are humourless.*' There was no denying that; no animal besides ourselves would be capable of laughing at the funny story above. Besides, that joke only works within our culture and our current era. Anyone who is unfamiliar with public conveniences with separate cubicles or who doesn't know how mobile phones work wouldn't get the joke at all. But the basic idea, namely a sequence that is interrupted by a moment of surprise, is universal and is regarded by many researchers as being the most important element in play. If one looks at playing humans or animals, it is clear that each game is characterized by a degree of creativity. It is the novelty or something unpredicted and surprising that makes a game attractive. If this dynamic is missing, then a game becomes boring, or a joke is rendered unfunny. Animals that play do so to simulate and train the broad spectrum of their behaviours. The fun and the pleasant sensations that we experience whenever we laugh at a good joke, play computer games or chess or football, or just racket about with our kids derive from the thrill of the new. Every new situation has something instructive about it, which trains our adaptability and ultimately our biological fitness. It follows that there must be mechanisms that drive us to invest so much time and energy in these modes of behaviour. A special issue of the renowned scientific journal *Biology* in 2015[4] on the topic of play in the animal kingdom pursued this basic idea. It therefore gives me great pleasure to be able not only to transport you to the world of playful apes and dogs but also to introduce you to ludic spiders, frolicsome birds and frisky Komodo dragons and fish.

The chemistry of enjoyment

One morning while I was researching this book, I was sitting at my
desk with my nose buried in the special issue of *Biology* mentioned
above. Veverin, one of my small sons, toddled up to me and sat on
my lap. Normally my screen is filled with text and of little interest
to him, but at this moment an image could be seen, showing a
crow perched on a roof holding something in its beak, and this
immediately drew my son's attention. *'What's that bird doing?'* he
asked. *'He's having fun on a sleigh-ride,'* I replied. And it was true:
Nathan Emery from Queen Mary University of London and Nicola
Clayton from the Psychology Department of the University of
Cambridge had used a private video from Russia[5] with more than
1.5 million hits on YouTube to illustrate their hypothesis about play
behaviour in birds.[6] In the video, a crow is seen sitting on a snow-
covered roof and using a white plastic ring as a toboggan. Looking
at this clip, you can't help but call to mind images of snowy pistes
with people screaming with delight as they zip down them riding
old inner tubes from car tyres. In another video, a pair of swans
surf on to a beach,[7] and you're reminded of people riding waves
on Hawaii. It's hard to credit, but looking at these videos, the two
researchers are getting close to answering an age-old question that
has long intrigued behavioural biologists: Why do we play?

Where young animals are concerned, this question seems easy
to answer. Animals – at least, those that have more complex brains
and which for the most part care intensively for their brood –
play in order to make themselves fit for their future lives. So, for
instance, animals train hard to perfect their hunting technique or
to attain as favourable a position as possible in the social hierarchy.
But where does their inner motivation come from? No wolf cub
playing rough-and-tumble with its siblings is conscious of the fact
it is doing so in order to be a better fighter in future or perhaps to

become pack leader one day. Also, the whole thesis about learning isn't wholly plausible anyway, for in that case why do many adult animals continue to play with great enthusiasm despite having already learned all the key modes of behaviour? A very similar enigma is birdsong; why do they sing throughout the season, when all it takes is just a couple of days in spring to pair off and mate?

Perhaps you can still recall the chapter 'Hormones in the driving seat' (see pp. 23–32), where I attempted to explain the effect that hormones have on our behaviour. There is a host of neurological mechanisms within our brains that control the most diverse messenger substances and whose receptors direct our behaviour. One example of this, which I will go into in greater detail in the section entitled 'Sentimentality' (see pp. 269–82), is our inbuilt system of rewards. The effective agents in this are the neurotransmitter dopamine along with endogenous opiates, which combine to put us in a pleasant mood. The reward system kicks in when we have mastered a biologically significant task. The feeling of elation that a predator might feel on capturing a prey item has nothing to do with stilling its hunger; it needs to eat before it experiences that. Instead, the sensation of having achieved something that involved a great deal of exertion, and in many cases also risks, is simply overwhelming.

Yet how does that work with motivation? Researchers in neurology and the learning process talk in terms of 'liking', 'wanting' and 'learning'. Hormones that feel good to us, and make us happy and contented, generate a 'liking' sensation. Because we desire to experience this feeling again (the aspect of 'wanting'), we repeat and practise those actions time and again, until we have picked them up ('learning').

Comparative neurophysiological studies have revealed that this reward system operates not only in us humans but also in animals.[8] Coming back to Emery and Clayton's study, the reward system has great significance for the behaviour of birds and for their singing. The great thing about this explanation is that the theory supplies plausible answers to a whole range of questions. Perhaps it may even apply to the almost absurd behaviour exhibited by wild field mice, which enjoy running around a hamster wheel. These fitness centres for wild animals were deemed nothing other than an amusement until studies demonstrated that wild mice were just as keen as laboratory-bred animals to spend a considerable amount of time on these treadmills.[9]

If we consider birdsong, then an obvious explanation suggests itself here too. Birds sing to impress females or to mark their territories. Now, as you have already seen, most birds are monogamous, meaning that they ought to have settled on their partner and territory sometime at the start of the breeding season. After a few days of loud calling in spring, therefore, the whole thing should be over and done with, and our world would fall silent once more. According to that same logic, animals should no longer play when they are grown up and have acquired their suite of behaviours. But in actual fact, many animals still play as adults and our delightful birdies go on singing throughout the season.

The mechanisms of dopamine and other neurotransmitters have been extensively studied not only in people but also in animal experiments with mammals, principally rats and mice. But what works for mammals does not necessarily apply to birds as well. Scientists have therefore hit upon the idea of suppressing dopamine in birds using drugs. If the theory about the reward system is true, then after a brief period of serenading, birds should get no further inner reward for their singing and should hence sing less than birds that had not been given any dopamine suppressant. Conversely, a supplement of dopamine should lead to more singing. And this is exactly what the experiment showed.[10] In other words, whenever you break into joyful song in the shower, or if you sing in a band or a choir, then in all likelihood you will feel exactly like the song thrush on my roof, since the neurological mechanisms are the same. Why should Mother Nature dream up something new when a tried and tested procedure functions admirably?

What is play?

Before we turn our attention to frisky reptiles, fish, insects and molluscs, we should first briefly acquaint ourselves with some terms from behavioural biology:
  What kinds of play are there?

- Locomotor play: Sequences of motion, like those of the afore-mentioned swans or that of a baby elephant playing in the surf, are rehearsed.[11]
- Object play: Physical objects are tinkered around with – say a kea (a New Zealand parrot species) nibbling on rubber auto parts.[12]
- Social play: This kind of play has to do with training animals in modes of social behaviour – for example, dogs frolicking around. For the most part, the objective is to defuse tensions and attain a harmonious coexistence within a group.

Further, what is play and what conditions must be fulfilled? Play behaviour must:

- be non-purposive in any given situation (for instance, not in order to achieve a particular social rank);
- be voluntary, spontaneous and enjoyable;
- be clearly demarcated from real behaviour or manifested in a phase of life in which it has no purpose;
- be repetitive without being stereotypical;

- occur without external stress.

Assuming you don't baulk at these biological definitions,[13] which incidentally can also be applied to human play behaviour, we will have some fun in the following paragraphs.

Animals react in very different ways when placed in front of a mirror. Later, in the chapter 'Who am I, and who exactly are you?' (see pp. 196–230), we will learn a lot more about the so-called mirror experiment. Most animals ignore it, whereas others go so far as to recognize themselves. Yet many animals regard their reflection as a social partner. Perhaps your biology teacher at school once organized an experiment of this nature. You take an aquarium housing a cichlid and dip a mirror into it. The fish immediately begins to defend its territory and rushes at the supposed rival. It does this to the point of exhaustion, so it is advisable only to conduct this experiment for a brief moment. Some biologists have been struck by the fact that these fish react aggressively not just when faced with a mirror but also to the presence of a thermometer in the tank. As you know, ornamental fish have to have their aquarium water at a particular temperature in order to thrive. Consequently, many aquarists put a thermometer in the tank, weighted at the base and with a small air bubble in the top so that it stands in an upright position.

It has been observed with white-spotted cichlids (*Tropheus duboisi*) in particular that they attack aquarium thermometers with gusto. The ingenious researchers even came up with a term for this behaviour: ATT ('Attacking The Thermometer'). In truth, though, this is not a question of genuine attacks of the kind that are carried out against reflections. Beforehand, using dummy fish, the researchers tried to provoke an aggressive reaction, and as a result were well aware at what point an object was seen by the fish as a genuine rival. According to the categories outlined above, the mock attacks on the thermometer were instances of 'object play'.[14] Employing the play criteria noted, this behaviour was unquestionably devoid of any immediate purpose, was voluntary and stress-free, had nothing to do with real behaviour, and was repeated frequently without being stereotypical. So, it is clear that fish also play. Although we still do not know what the neurological bases for this are, this insight will surely not be long in coming.

It is a similar story with the African softshell turtle (*Trionyx triunguis*), which behaves in similar fashion, playing enthusiastically with coloured rings,[15] while Komodo dragons play with various objects – for example, a piece of horsehide.[16] Gordon Burghardt, Professor of Psychology at the University of Tennessee at Knoxville, put forward the idea of watching videos of Komodo dragons speeded up, because, according to him, viewed in this way their playful behaviour bore an uncanny resemblance to that of dogs. I discovered a video on YouTube showing a Cuban rock iguana (*Cyclura nubila*) greeting its owner with evident delight, and I would somehow not have been at all surprised if it had gone on to bark and wag its tail.[17] Even so, I did find myself wondering why a species threatened with extinction was being kept as a pet in the first place.

The giant Pinta Island tortoise of the species *Chelonoidis abingdonii* was wont to greet its keeper at the Charles Darwin Research Station on the island of Santa Cruz in the Galápagos in a very similar manner. The last surviving individual of this subspecies, which the BBC dubbed 'Lonesome George', describing it as the loneliest animal on the planet, died in 2012 at the age of just over 100 and weighing some 90 kilos (200 pounds). Giant Galápagos tortoises grow throughout their lives and can reach weights of over 400 kilograms (880 pounds). It is a great pity that we have been deprived of this species for ever, simply because it tasted so good.

But let's put this sad tale to one side for now; after all, the matter in hand is play. It will presumably come as no surprise if I tell you that some representatives of the mollusc phylum, namely certain species of octopus, are inclined towards playfulness. New objects in particular are investigated inquisitively at first, before being extensively played with.[18] This too is a case of object play, which is hardly surprising since these octopuses neither live in social communities nor practise brooding care.

However, I do find the following two examples absolutely astonishing: young paper wasps of the species *Polistes dominulus* practise ranking fights some months before the actual contests take place to decide the hierarchy,[19] and the American spider species *Anelosimus studiosus* practises copulation with its sexual partner while still a juvenile, despite the fact that this cannot possibly result in fertilization since the animals are not at that stage sufficiently sexually mature.[20]

The list of unbelievable things is not yet exhausted, however. Remember the experiment with the hamster wheel? The treadmills were simply set up in the wild for use by any animal that came upon them. To the complete amazement of the researchers, frogs started working out on them, and even – and this will stretch the bounds of your credulity – snails.[21]

If we work from the assumption that playful exercise enhances general fitness, it would hardly be remarkable that such a behavioural strategy has evolved across different species and systematic groups quite independently of one another. According to the 'surplus resource theory', all it takes is the availability of sufficient food, time and security for playful behaviour to arise as if by chance. Of course, as it stands it is still completely unclear whether this behaviour is also intensified by self-reinforcing mechanisms such as the reward system among mammals and birds.

★

Anyone who breaks the rules loses!

Let's look now at a mode of play behaviour that every dog owner at least would have something to say about. If you do keep a dog as a pet, you'll be familiar with the following situation: you're on a walk and your dog isn't on a lead; instead, he's running a bit ahead of you and then doubling back, sniffing scents here and there. Suddenly another dog appears on the path in front of you. There's no sign of its master or mistress. Both you and your dog take a long, hard look at the newcomer. You're studying its body language for signs of aggression, indifference or readiness to interact in a friendly fashion. To your great delight, the dog lies down with its front paws outstretched, raises its haunches and wags its tail. Your dog is even more pleased, as he's not going to let pass such a friendly invitation to frolic. So he rushes up to the other dog, bounding high into the air. At this point, the dog's owner appears round the corner and gapes wide-eyed in shock at the scene; he hasn't been witness to the communication that passed between the dogs, and so he's afraid of what might befall his faithful companion. But a split second later his tense expression relaxes once more, replaced by a broad grin of pleasure when he sees it's all just a game.

Once you have had a really good look at the body language of dogs, it's not at all difficult to understand. Marc Bekoff, Emeritus Professor at the University of Colorado, is possibly the world's most eminent expert on dog behaviour. He has evaluated and 'translated' countless interactions with both dogs and coyotes. He interprets certain body postures as 'I'll bite you now, but only in fun' or 'Sorry I overstepped the mark there; I didn't really mean it'. He believes he has identified four fixed rules in dogs' play behaviour:[22]

- Ask before you play.
- Be honest.

- Stick to the rules.
- Apologize for any misbehaviour.

In the course of Bekoff's studies, it emerged that these rules are almost always observed. There were transgressions in only 2 to 5 per cent of the interactions. From my personal experience I can concur with this, though this only applies to dogs that have been socialized. Dogs also have to learn the rules of social interaction with one another. Although it's not a question of a culture here but rather behaviour that has been genetically preprogrammed – in other words, predetermined – that too can only be learned in appropriate situations. If it is never acquired, the dogs do not behave normally or do not stick to the rules. For instance, among coyotes, which are close relatives of dogs and wolves, it has been observed that animals that do not observe the rules tend to leave the pack and that this social exclusion leads to reduced fitness and premature death.[23]

Naturally, our domestic dogs do not run this risk, as they are provided with food and shelter. If they fail to behave socially, we put them on their lead and so prevent an escalation and serious bite injuries. Undoubtedly, there's a divergence of opinion on the question of whether dogs should be kept on leads, and I do respect the fear that many people have. Nevertheless, we should be clear that where legislation compelling owners to leash their dogs has been introduced, this has over time changed the behaviour of our canine companions. Dogs that are constantly kept on a lead are incapable of ever developing appropriate social behaviour towards other dogs. They have never had the opportunity to learn the ground rules of peaceful coexistence and so are in the truest sense of the word asocial. Even though most dog owners keep their dogs on a leash because they think it is a way of protecting them and also expect other dog owners to reciprocate, from the point of view of behavioural biology this is fundamentally wrong. Cause and effect have been turned on their heads here: the dogs behave in an asocial manner because we keep them on a lead. One sad side effect of this is the impoverishment of the social lives of the dog owners themselves. Before leash laws, people would stop and chat, whereas all they do now is rush past one another as quickly as possible, dragging their choking dogs along behind them.

But if we don't keep our four-legged friends leashed, we are treated to a masterclass in sociability. In the normal course of events, domestic dogs that have gone feral and their closest relatives the wolves live in relatively stable social groups, namely their packs. They react in an unfriendly way towards other packs. Our domestic dogs' packs consist of the human families that own them. Whenever a dog encounters another dog on a walk, from an evolutionary standpoint this represents a major challenge. It is something of a miracle that, in the vast majority of cases, these encounters unfold playfully and peacefully. This is down to the aforementioned body language, certainly, but more precisely to a shared lifestyle that has been in existence for more than 200,000 years.[24] Our dogs have developed a very keen sense for the context of any given encounter. If people greet one another in a friendly fashion, then it is only natural for our dogs to act in a similarly peaceable way. Perhaps we might even be justified in this instance in speaking in terms of a genuine inter-species understanding based on reciprocity. Of course, these encounters only pass off peacefully where normally socialized dogs are concerned, and the following example makes clear how vital the act of socializing is in the early development of an individual character.

One of the most extensively studied social groups among animals is that of the yellow-bellied marmots (*Marmota flaviventris*) of Colorado, which virtually live on the doorstep of the Rocky Mountain Biological Laboratory of the University of California. This group was first studied as far back as 1962. It is a real delight to observe the antics of these amusing little rodents, and so it is perhaps no surprise that the researchers were patient enough to log almost 30,000 social interactions among these animals over a period of ten years. The findings that emerged constituted one of the most substantive pieces of evidence underlining the significance of playful confrontations between mammals.[25] The researchers managed to demonstrate that juvenile marmots that emerged victorious from playful interactions went on to occupy a dominant position within the group. Perhaps you are now thinking that you could have guessed that anyway; however, the remarkable thing about this study is that the hierarchy among adult marmots is established without any dramatic ranking fights. Thus, the hierarchy is already generated in the course of individual development. The advantage of this is obvious, since this is nature's way of ensuring that a hierarchical order evolves without the necessity for serious injuries to be sustained.

We humans also practise our social behaviour through play and in so doing develop strategies that preclude our having to constantly fight vicious battles. For instance, we send our alpha males and females, such as senior politicians, managers and team leaders, to outdoor camps for them to learn sensible social behavioural strategies like team-building. One key aspect is developing trust. Bosses need to learn that they can trust their colleagues and that they don't have to do everything themselves. In this kind of exercise, therefore, a participant is blindfolded and a rope bound round his or her ankles. The rope runs over a pulley suspended in a tree, and the alpha individual is hoisted up feet first and hangs there for a short while, swinging to and fro. All the while, they are held suspended by their comrades and have to trust that they won't suffer a collective sneezing fit and let them plummet to earth.

Juvenile bonobos are trained in much the same way as our

elites. In the 'hang game', a stronger individual holds another by the ankles and dangles it from a tree. This trust-building measure is without a doubt an incredibly strong social bonding agent and presumably works so well because it's simply a whole lot of fun when the world is topsy-turvy.

There are countless other observations of this kind, and I am most grateful to the researcher Isabel Behncke, who has intensively studied social behaviour in general and that of bonobos in particular,[26] for posting her videos in a playlist on YouTube.[27]

What's special about bonobos' play is their great creativity. The renowned primatologist Frans de Waal described one of these games as 'blindman's buff'.[28] This involves a female bonobo covering her eyes with a banana leaf and stumbling around on a climbing frame or deliberately bumping into other animals, which take great delight in this.

If anyone should think that humour is an exclusively human preserve, I recommend that they watch a video put out by the US broadcaster CBS News. It's a report on a small orangutan from Barcelona Zoo. A young man is seen performing a simple magic trick involving a chestnut in a cup – a bit like the 'shell game' – in front of the animal's cage, and the ape is quite literally bowled over by it.[29] Although I am totally against keeping great apes in zoos, I still found this video really moving, since the young orangutan is clearly so delighted that all further comment is extraneous. Little wonder, given that the capacity for laughter evolved among anthropoids some 10 to 16 million years ago, according to genetic studies.[30] On the other hand, if you ever hear a Eurasian magpie[31] laugh, you might be tempted to think that our common ancestors the dinosaurs already had something to laugh about even in the Jurassic period.

*

15. The Oedipus complex

The new generation of millennials is so mollycoddled and lacking in independence, don't you think? All today's youth does is stick around the Hotel of Mum and Dad enjoying free board and lodging until they're almost the age when they could be grandparents themselves, and if they ever do move out it's no time before they inherit the house anyway. But leaving aside the financial and social pressures that are driving this behaviour, which have been debated long and hard, adopting this social strategy is actually quite a smart move. The resolution of intergenerational conflict and more liberal attitudes to bringing up children make it a more attractive prospect nowadays for our offspring to remain at home longer than used to be the case. It's not simply a question of getting their underwear washed or having their breakfast served up to them with no effort; rather, it's often also to do with the fact that parents and children nowadays tend to regard one another as partners offering mutual support. Perhaps we should be less critical of this development and see it instead as a compliment to both generations and throw overboard once and for all Freud's notorious idea of the Oedipus complex.

Yet it is remarkable that young women tend to fly the coop earlier, at an average age of twenty-two, than young men, at twenty-six. It could be that this behaviour is deeply rooted in our anthropoid ape culture, for the situation is very similar among our closest relatives. The young females leave the group and search out a new social network for themselves, whereas the males remain with their mothers and wait for the young females from neighbouring groups to come and pay court to them. This is a considerable challenge for the young females, for naturally resources are limited and groups are reluctant to share them with strange interlopers. The only thing that counts in their favour is their sex appeal and the ability to turn the heads of the males. Once they have achieved this, the alpha males in the group make every effort to ensure that the objects of their desire are allowed to stay. In this historic context, it comes as no surprise that human sons and heirs were also traditionally philopatric and the women they married, no matter whether they were a princess or a peasant girl, were obliged to move into the new household.

Once a female has succeeded in getting herself accepted into a group, then – according to the structure of the individual group – she is free to mate with one or even several of the dominant males. Among bonobos, even males that are less dominant sometimes get a look-in. In this, they get robust support from their mothers. Now, it may not be to everyone's taste to get down to the nitty-gritty in the presence of your mother, but if it helps, what the hell…

Bonobo females enjoy a higher status in their groups and can even separate aggressive males without putting themselves in danger. Clearly more dominant bonobo males show similar respect in cases where an impudent juvenile male, accompanied by his mother, proceeds to have his way with an attractive female.[32] But if the mother isn't present, then the fur flies. For this reason alone, it is a distinct advantage for young males to stay with their mothers.

However, there are other advantages to remaining under the care and protection of one's parent for longer. We acquire most of our knowledge in childhood. It is therefore unsurprising that those animals that are more highly evolved socially put a premium on protracted care of their offspring. But only a very few species of animal live for long following the menopause – the great majority die shortly after ceasing to reproduce. We humans are a major exception to this, since we can go on living for many decades after the menopause and (in the case of men) the andropause. The reason for this is obvious: we have a major stake in passing on knowledge to both the next generation and the one after that.

Thus far, we know of only one other vertebrate with a comparably long menopause: female orcas. On average, they cease to reproduce at around the age of forty, but can live to ninety. Ongoing studies involving the analysis of data collected over thirty-six years of observation have yielded some quite astonishing results. It turns out that male orcas are real 'mummy's boys'. The studies showed that even some thirty-year-old orcas were still dependent upon their mothers. In addition, it emerged that the probability of such orcas dying within a year was fourteen times higher if their mothers perished or stopped looking after them. Female orcas are not so hard hit by the death of their mothers, but even so the likelihood of them also dying sooner is still three times higher in that eventuality.[33]

These results impressively demonstrate how long even adult animals remain dependent upon their mothers. Considering this finding in conjunction with the stability of orca culture, we humans should respectfully take our hats off to these creatures.

16. A monarchy with room for democrats

Relax – I'm not about to bore you with tales of ants and bees, which as colony-dwelling ('eusocial') insects have a queen, but which follow quite different rules to us mammals. But perhaps I will be able to surprise you by telling you that it's not just among us humans that social position is determined by accident of birth.

We have learned that it is customary among our nearest relatives for young females to leave their own social group and to go off in search of a new group. But the converse is also the case. Among spotted hyenas, it is the juvenile males that must make themselves scarce. You should also be aware that the social groups within a spotted hyena clan are stronger than in a wolf pack.[34] But the males are not responsible for this, for the 'movers and shakers' in these groups, each of which comprises some eighty individuals, are the females. Even a female that is on the lowest rung of the hierarchy takes precedence over the most dominant male. But even the females do not have an easy time of it, as they are constantly required to assert their position in the hierarchy by displaying aggressive behaviour. Yet despite this, some animals still have it easier than others, since the mother's social position within the hierarchy is inherited.[35] Just like in a monarchy.

But hold on, I hear you say, it isn't just in royal families that social position is inherited. Within normal societies too, a clear correlation exists between the social status of parents and that of their children. A better education, a higher income and a good network ensure that some children enjoy better prospects of advancement than others without having to exert themselves. Of course, that isn't fair, but it is natural. Animals with a long rearing period are given every conceivable support by their parents. This ensures that one's own genes can be successfully passed on by one's progeny. The scientific terms for this are nepotism or favouritism.

In spotted hyenas, social position is even reflected in their genes. Better placed animals have been found to have longer telomeres.[36] Telomeres are appendages at the end of our DNA strands, which are shortened with every cell division. Once they have disappeared, the cells can no longer divide and so they die. This mechanism determines – among us humans too – how long an individual lives, and hence also indirectly its physical fitness and reproductive success. The trick with the telomeres is just one of several mechanisms that nature has come up with to kill us, for it has no interest in us living for a long time. On the contrary, where nature is concerned, once we have successfully reproduced, we can and should take our leave as quickly as possible. This makes perfect sense from an evolutionary perspective, since our offspring are altered or mutated ever so slightly. This mutation makes them either better or worse, and the mechanism of selection ensures that only the best survive. In the absence of this mechanism and the genetically programmed death of the parents, there would be no evolution. But as individuals too, we should be grateful to telomeres, since by inhibiting the process of cell division they also protect us from cancer.

But coming back to the spotted hyenas, a corrective should be offered in their defence at this point. In contrast to their relatives the striped hyenas and brown hyenas, spotted hyenas are not carrion-eaters, as was wrongly claimed in many older documentaries. On the contrary, they are accomplished hunters that can reach a top speed of up to 60 kilometres (37 miles) an hour. By means of skilled teamwork, they generally prey upon animals that are much larger than themselves. It is also a myth that they steal prey items from lions. The opposite is in fact the case, for the king of the animals has an ear that is finely attuned to the communication sounds of the spotted hyena, and can pick up from a great distance the noises they make when feasting. It then makes its way at a leisurely pace to the 'larder' and proceeds to eat its fill of the spotted hyenas' hard-won prize.

Among baboons too, juvenile animals benefit from their parents' higher social standing. For many people in Africa, the behaviour of troops of baboons is often very disagreeable, since they have no fear of humans and are a typically synanthropic species. They besiege busy roads and encroach on settlements to try to snatch scraps of food. These animals can weigh as much as 30 kilos (66 pounds) and have an impressive array of teeth, and anyone with a fear of dogs would be well advised to give baboons a wide berth.

Baboons live in groups, and within these there must be one dominant animal that takes key decisions: how long to sleep, where the troop should go on any particular day, how long to rest for, whether a location is safe, and where to make camp for the night. So, if one could find out why baboons come to the decision to help themselves to humans' food, and discover which individual dictates this, it might be possible to influence the baboons' actions in a way that encourages them to modify their behaviour. In any event, that would be preferable to shooting or poisoning the animals. In theory, those options are open to people, since baboons are not currently a threatened species and so are not protected by international conservation conventions.

A range of important findings on decision-making in social baboon groups emerged from one such research project, and was published in *Science*, one of the foremost scientific journals.[37] Basically, there are only three possible ways of reaching decisions within a social group:

A: Despotically
B: Hierarchically
C: Democratically

Although there are frequently several dominant males within the group, one is indisputably the boss. Yet there are also certain females that rank very high up in the hierarchy. As a result, it is a very complicated business studying which baboon makes decisions for the group, and when and why this happens. The researchers therefore fitted every animal with a satellite tracking device, which enabled them to pinpoint the exact positions of each of them over a particular period. In this way, they were able to determine exactly who was with whom at any given time and who followed whom most often.

It became clear that many of the baboons were following the individuals numbered 10, 17 and 20. These animals were therefore higher up in hierarchy and others followed them. On the other hand, there were animals like number 3, which not a single baboon followed throughout the entire period of observation. Yet what surprised the researchers the most was animal number 17. As the arrows clearly indicated, lots of baboons followed this animal, which led them to believe that it must be old, experienced and strong. But in actual fact it turned out to be a juvenile, which brings us back to the question of heirs apparent. Of course, in turn this young animal followed its mother, number 20, and consequently all the baboons that followed the juvenile number 17 were ultimately following number 20 too. This observation clearly demonstrated what a strong influence the social status of one's parents exerts.

This was not the only finding from this study, though. The situation just described referred to an average of all data across the entire duration of the observation. But on any given day, important decisions were not being made every minute, so the key thing was to look at the animals' behaviour in those moments when something actually happened. It quickly became evident that the animals were not following a 'despot' that was directing the fate of the whole troop by virtue of its dominance. Hence the only other explanations were that decisions were being arrived at hierarchically or democratically. And it's here that things get interesting. It's likely most people (myself included) would guess that the decision-making happened in a hierarchical fashion. In fact, the opposite was the case. If one looks at the specific moment in which decisions affecting the whole group were made, what unfolded was a dynamic process, in which some animals did one thing and others another. Then more joined the latter group, leading those in the former to change their minds and throw in their lot with the majority. At the end of the day, the decision-making process was a democratic one. Of course, there is now much speculation as to what extent a democratically organized social life is advantageous in an evolutionary context. No need to worry, though – there's no question of us humans being upstaged here, for even if baboons live in a democracy, they don't have a word for it.

It is common knowledge that democracy is a useful tool where decision-making is concerned. But did you know that the transmission of cultural assets from one generation to the next is also subject to democratic mechanisms? A study conducted on humans, chimps and orangutans investigated the criteria by which they learn from others or acquire elements that then become part of a cultural heritage. It also transpired that an action was learned more readily if it was performed en masse rather than by just a handful of individuals. In the experiment, this meant that either an individual needed to perform the same act three times or that the same action was carried out just once each by three different individuals. This ingenious, neat experiment in behavioural biology came to the conclusion that chimpanzees are just as democratic as we are, whereas orangutans are not.[38] This has its origins in their modes of social behaviour, which we will examine presently.

Experiments like this are not conducted for their own sake, but are meant to tell us more about human behaviour. Have you ever asked yourself why there's such a glut of well-known inferior art around with a large advertising budget and, conversely, so much obscure good art with a small advertising budget? Or why something that's obviously correct is thought by most people to be untrue? Or why a riverside park in Munich that's open all year round is suddenly visited by over 6 million people in the space of just a couple of weeks? Whereas if just one person were to enter the park every hour for a period of 1,000 years, presumably no one would raise a peep about it. The logic behind this behaviour is simple: it is safer to do something that lots of others are doing than to emulate a solitary individual who practises his behaviour in constant repetition.

But let's return to the question of our inherited social position.

Just as one can be fortunate in one's birth, equally one can experience bad luck, and sometimes the sex you are born into means you have been dealt a bad hand. Rhesus monkey mothers treat their sons so meanly that no close-knit social bond ever develops between them and the young males consequently leave the community very early.[39] This is a harsh fate for the youngsters involved, but it's good for genetic diversity.

17. Bestial biographies

What an idiotic chapter title, you may well be thinking. After all, animals live in the here and now and don't possess any long-term memory. If they have any at all, it is very short term. How can they have a biography, then? A biography of animals written by humans would be a so-called CITES (Convention on International Trade in Endangered Species) Report.[40] This records when and where a particular animal was trapped and how it was transported to this or that zoo. Zoos then also keep information on individual animals such as diseases and illnesses contracted, regular health-check records and details of births, along with family relationship data. But I mean *actual* biographies of animals – animals that have a lifelong memory, that have learned from their experiences and are also capable of thinking of, and planning for, the future.

During one of my first visits to a zoo that I can remember, a small baby elephant was led out of its enclosure. It came straight up to me and my parents explained that it was an elephant. For me as a small child, who had to crane my neck to look up at this 'baby', the word 'elephant' imprinted itself deep in my memory. My parents then proceeded to tell me the story of an elephant that used to travel around the country with a circus. One day, when the elephant was still very young, a baker came out of his shop and gave the animal a bun. Many years later, when the baker had long since passed away, the circus visited the town once more. The elephant stopped dead in front of the shop and refused to budge. Then the baker's son recalled an incident in his childhood, when his father had given an elephant a bun, and promptly did the same. To everyone's relief the elephant then moved on, cheerfully munching away. I can't say whether this incident really happened or if it was apocryphal, but many people I have met have heard a similar story at some time or another. The elephant is credited, reverentially but still with a degree of scepticism, with a good memory, and people are inclined to believe that this one remembered where he'd been given food. Is such a thing really possible? Do animals exist that can recall events that took place decades in the past, and which on the basis of these experiences can make decisions relating to the present?

This is a tricky business, even for us humans. When I'm chatting with my wife about a holiday in Greece that we went on when we were students, I find the stories we recount to one another are quite different. This is due to the fact that we do not have a photographic memory, but rather store occurrences in engrams, namely neural networks. In order to retrieve these engrams, we require a key stimulus. This can be a smell, a similar environment or even a piece of abstract information like the phrase 'holiday in Greece'. The neurological equivalent of this key stimulus races through our brain, searching for a comparable pattern. Once located, our brain attempts to reconstruct the information that has been stored as an engram. But even though two people have experienced exactly the same thing, their engrams do not resemble one another in the slightest. Each engram fits into the existing network and in turn connects with others. The problems begin when two people start comparing their recollections.

Yet how can it be that our memories are so drastically different? Naturally enough, researchers have also busied themselves with this question, and accordingly have tried setting traps for our memory. One of the best-known of these traps is the 'lost in the mall' experiment. In this, the test subject is presented with a series of anecdotes that close relatives have written down about them. One of these stories involves our test subject getting lost in a shopping mall as a kid and being found and brought back by a stranger. The subject is then asked to fill in some further details about this incident from their own recollections. A quarter of participants were able to flesh out the situation, describing the person who helped them, or giving the name of the mall and even those of individual stores there. The catch is that the anecdote was a pure invention. So, in the case of a quarter of the test subjects, their memories were manipulated and they experienced a genuine recollection of an incident that never actually occurred.[41]

Of course, manipulations like this can also happen in real life, and our brain is forever making these kinds of mistakes. As such, it is hardly surprising that two people should have a totally different recollection of a particular event. In the general run of things and among friends, this all just makes for greater amusement. But what if a daughter remembers being raped by her father, or if a perpetrator identified in a police line-up is convicted even though they are innocent? The psychology professor Elizabeth Loftus[42] is credited with having discovered the phenomenon of the malleability of human memory, and her publications on the subject have earned her public opprobrium and even indictments.

Most people regard their memory as being infallible. What's more, a person defines his or her own self through memory. We see ourselves as the product of our experiences and ask ourselves who we would be if we lost our memories. Who would I be without my experiences at school, without my recollection of my first disappointment in love or the indescribable joy I felt at the birth of my children? We are nothing but the sum total of our episodic memories. And then someone comes along and informs me that I've simply fabricated large chunks of my memory!

Perhaps you're thinking: that's all very well, but surely he just means the trivial stuff here... and I have a crystal-clear recollection of the things that really matter. Sadly, the exact opposite appears to be the case. Precisely with things that are likely to make the greatest impact upon us or even to traumatize us, the workings of our brain are far from perfect. This factor is now familiar to every lawyer who has to deal with witness statements, and no matter how confident and plausible a witness might be, his or her neurological foundations are not.

But can a person's memory also be influenced in a positive way? The answer is yes. Elizabeth Loftus devised a little experiment along these lines. Her test subjects were told that they had suffered from a metabolic disorder when they were small children. This had been caused or exacerbated by overconsumption of chocolate, ice cream and other sweet things. In the course of a meal that was subsequently laid on, researchers analysed the choices the test subjects made at a buffet bar, depending on whether they were affected by false memory or not. Presumably you won't be surprised to learn that those with a false memory went for the more healthy options.[43] But was it wrong to use a little white lie concerning childhood memories to protect people from obesity and all the associated health risks?

What's more, it has recently been shown that the hippocampus (an area of the brain associated with learning) is negatively influenced by obesity and that those affected are slower at learning[44] and hence cannot record so many memories. The trade-off here was therefore one little false memory in return for a multitude of genuine ones! Elizabeth Loftus likes to compare her procedure with the story of Father Christmas, which we have no problem telling our children because it enriches childhood and we all have wonderful memories of this happy time. To be quite honest with you, I can't decide on the moral integrity of this.

But what does all this have to do with the memory of our circus elephant? To answer this question, we need first to look at the two forms of long-term memory that we have at our disposal. One of these forms of memory is not so easy to fool as the other. To all practical intents and purposes, our procedural long-term memory is immune to any influence whatsoever. Once you have learned how to ride a bike, walk or swim, then you can rely absolutely on never suddenly breaking into swimming motions while walking. I'm also assuming you'd have no difficulty in ascribing this form of long-term memory to animals too. Then there's declarative (or explicit) long-term memory. This form subdivides into episodic and semantic memory. The latter encompasses our knowledge about the world and enables us to understand what's meant by the terms 'Greece' and 'holiday'.

I'm sure you can all think of films and books in which a character suffers amnesia. Our unfortunate heroes can't remember any personal details, but they are whizz-kids with computers or have retained some other skills from their former professional lives. With this form of amnesia, our semantic memory is not affected. An architect can still design houses, a surgeon can still operate and a teacher can still teach. What's affected is the most sensitive and, as we have seen, the most easily influenced form of memory, namely our episodic long-term memory. This is where we effectively store ourselves, our pasts and our visions of the future, as well as everything that has made us what we are. It is all the more dramatic that precisely this form of memory should be so sensitive and easy manipulable.

The question now is: do animals also possess this form of memory? If that were the case, then one ought to be able to manipulate it and thereby influence their behaviour.

The most popular animals for use in this kind of research are not our nearest relatives, the great apes, but mice. In genetic terms, they are not nearly so close to us, but they are cheap and don't live very long. Plus, in the case of mice virtually no difficulties arise with ethics committees in granting permission to undertake animal experimentation, nor is there any danger of a public outcry. Of course, researchers do not undertake experiments on memory in mice in order to discover how a *mouse's* memory functions. No such proposal would secure any funding. But if the experiments on mice are able to help us understand the human brain, then it seems the funding possibilities are almost unlimited. After all, what's at stake here is nothing less than the loss of people's identity and the attempt to preserve episodic memory in Alzheimer's patients.[45] Who knows, perhaps scientists will even one day come up with a pill to promote better learning. In other words, there's an awful lot of money involved here. One of these experiments actually succeeded in implanting a false recollection into the episodic memory of a mouse. The mouse's behaviour in a particular context gave clear evidence of this recollection, when it suddenly, in a familiar and unthreatening situation, reacted with panic and 'froze'.[46]

In order to make this experiment comprehensible, I need to grossly simplify something that is extremely complex and draw a somewhat dodgy analogy. You've all heard of the concept of formatting a hard disk, I'm sure. In principle, the moment that occurs, nothing is really being erased. But all of a sudden the hard disk's index goes missing. The data telling us where files are located have been erased. In the end, then, it is only the signposts that have been uninstalled, whereas the destination still remains in place. In our brains, the hippocampus assumes this function; it is rather like a signpost to the individual fragments of memory, since it knows where it has stored which piece of information.

The hippocampus is also involved in the development of

epilepsy, and it was for this reason, in 1953, that doctors in Hartford, Connecticut, surgically removed large parts of the hippocampus from an epileptic patient by the name of Henry Gustav Molaison. The operation was successful, and Molaison's epilepsy was cured, but the downside was that he was henceforth incapable of learning anything new and had only limited access to his memories. The hippocampus is therefore an extremely important part of the brain, and was invented by nature at a relatively early stage of evolution.

The method employed by the researchers in our mouse experiment was nothing short of a master stroke. They drilled into the skulls of the mice with a 0.5-millimetre drill and infected them with a virus that smuggled new genes for protein synthesis into some particular cells of the hippocampus. These new proteins intervened in the function of the cells and could be directed by light. Since it is extremely dark in the middle of the brain, a fibre optic cable was implanted into the operation hole. This made it possible to switch the function of the manipulated hippocampus cells on and off, or to send the wrong signal. In this way, the researchers were able to switch on an engram that triggered fear in situations in which there was normally no reason for concern. The mice, finding themselves reminded of a situation in which they felt under threat, behaved accordingly. At the end of the experiment, their brains were removed and studied further.[47]

You might be asking yourselves why I have explained this procedure in such detail. Ultimately, my intention was to show that animals other than ourselves also possess an episodic long-term memory and that even these creatures are capable of recollecting episodes in their lives that influenced their behaviour. If that were not the case, there would be no point in conducting experiments such as the one manipulating the memory of mice.

*

As we have already indicated, it is possible to approach the problem in another way. Behavioural biologists typically observe the behaviour of animals and from this make inferences about the cognitive abilities underlying it. We have already seen that elephants live in a culture. Because a culture is all about reciprocal learning and changing one's behaviour accordingly, it is only logical to presume that lifelong memory is a factor in this. But I would like to further reinforce this point. One clear finding to emerge from studying elephants was that herds that were led by an experienced mother animal stood a better chance of surviving periods of extreme drought.[48] By contrast, in groups that were not led by an experienced animal, juvenile mortality was exceptionally high.[49] In times of drought, the few experienced animals that had not been slaughtered were able to locate additional sources of water and food because they could remember them. When all's said and done, this is no different from our elephant standing outside the bakery – thus making this story entirely plausible.

Another striking example is provided by dolphins. In their early youth, they develop an individual whistle. Every dolphin has its own whistle, and it is not at all far-fetched to draw a comparison with our names (more on this in 'Personality' in the chapter 'Who am I, and who exactly are you?'; see pp. 196–230). Obviously, you can remember the names of other people for decades. Through underwater loudspeakers, the researchers played the dolphins recordings of the 'names' of other dolphins. The dolphins made clear their recognition of some of these names – perhaps the animals in question had at some stage been kept in the same dolphinarium as the dolphins whose whistles they were listening to – with unambiguous responses. In this way, it was possible to show that these animals have a memory going back at least twenty years.[50] That is remarkable, since only very few animals in dolphinariums ever reach such an age. I cannot imagine, nor do I wish to, what a dolphin mother would have felt on hearing the voice of her child that had disappeared ten years previously. Besides, Whale and Dolphin Conservation (WDC), the world's largest international cetacean conservation organization, with whom I worked for ten years as scientific head of the German office, has spoken out against such experiments. We simply cannot predict what psychological consequences may result from them.

However, it is not just mammals that have a good memory. Like squirrels, ravens lay in stocks of food for the winter. Unfortunately, despite being mammals, squirrels do not have good memories and often fail to relocate their stores of nuts. Happily, though, many squirrels lay in winter reserves, and so the animals' noses give their poor memories a helping hand and the food stash of a fellow squirrel becomes a vital lifeline.

It's a different story with the ravens. They memorize their hiding places in sequence. This 'serial learning', as my doctoral supervisor Professor Todt dubbed it, is an ancient and very effective invention of Mother Nature. Like the memory exercises familiar to us humans, it involves learning a sequence, thereby lending wholly disparate elements or concepts a context that one can easily memorize. The ravens not only make a mental note of the place where they hide food, they also have a conception of time. They even know when their food reserves are ready to eat or about to rot.[51]

Among the memory artists of nature, the honey bees are a surprise package. Studies of them also suggest they have a very simple form of episodic memory or something similar. They can remember when and where a food source becomes available. Accordingly, they do not simply fly about haphazardly but make straight for the spot where, in the past and under given conditions such as a particular time of day, their search was amply rewarded.[52] It would once have been thought impossible for a simple ventral nerve cord system of the type possessed by honey bees to pull off such a feat of memory.

Alongside the ability to clearly recall hiding places and food sources, an episodic memory also has inestimable advantages in social life. By making a mental journey back in time, I can delve into my own past, and remember whether a certain person got on my nerves or was my friend. An episodic memory is therefore an essential prerequisite for a social life where it all comes down to individuals.

18. Like Facebook, only different

Love it or loathe it, Facebook was an invention just waiting to happen. The human face is worth far more than any identity card, since it makes us easily identifiable not just to other people. For the most part, inscribed into our face is also our past, our present and even our future. We can read a great deal from faces, and that is precisely why Facebook has been such a rip-roaring success. If we hadn't invented computers in order to conduct our wars more efficiently, we'd surely have invented them to manage our complex social lives. At least, that would certainly fit with the trend towards a more aggregated culture, since keeping in touch with friends, friends of friends, relatives and the people they know, colleagues, colleagues of colleagues, and holiday acquaintances is scarcely imaginable any more without electronic assistance. Yet we're not alone in having these problems, for some animals also have friends and relatives and inhabit complex social networks. Perhaps the most complex animal network identified to date was discovered off the coast of Western Australia.

As a biologist, one occasionally enjoys advantages that are denied to other people. Traversing a natural landscape, strolling in a city park or swimming in the ocean, you're moving around a world that is imperceptible to most of your fellow human beings. Using what is almost a kind of X-ray vision, you look into the canopy of trees, see rain falling on the foliage and sense the moisture transpiring through the pores of leaves. And as a behavioural biologist, one sees much more in zoos than the other visitors, and when you're standing in still water in your bare feet, it's like being an art lover in the Louvre. Just as an aficionado of painting can stand in front of a canvas and see not just the beauty of the finished picture but also appreciate how it was executed, and call up an image of the artist and the period he or she worked in, so a biologist can look at the sea and know what's going on down there. That's pretty much how I felt some years ago as I stood on Monkey Mia beach on Shark Bay and gazed at the water. Though I didn't see a single dolphin, I knew for sure what was hidden beneath the shimmering surface: the most spectacular social network so far discovered in the animal kingdom, as already discussed in the chapter 'Gangbangs' (see pp. 19–23).

The question remains, though, as to how to recognize an animal network. Most people will concede that animals have social lives too. We know that wolves live in packs and that cows gather in herds. Furthermore, there are shoals of fish and swarms of insects. But do these collective terms really characterize a social life? Strictly speaking, all they do is describe the outward manifestation of the phenomenon.

If you really want to study social life in the animal kingdom, then you have to focus on individual animals. These animals are sometimes in the company of one member of their own species, and at other times with another. If you spend a long time closely observing who gets together with whom, you begin to recognize alliances, which biologists have determined fall into three different orders.

Imagine you're sitting over breakfast with your family on a Sunday morning. You're chatting about the week just past, and making plans for the one ahead. At this moment, you find yourself in your first-order alliance. All animals that live in family groups inhabit this order of alliance. This form of coexistence can be very easily explained from a behavioural biology perspective. Each separate individual that invests in a community also invests in its own genes. However, among the great apes (including us humans), as well as among elephants and dolphins, there are also same-sex groups that form first-order alliances. Male dolphins frequently stay together with one or sometimes two male partners for their entire lives. Their identifying whistles, in other words their 'names', have even been seen to coalesce into one (see the section on 'Personality', pp. 204–17). This is a quite incredible development; as with our surnames, one 'name' becomes the identifier for two distinct individuals. Inevitably, family groups and intensive cooperation are rewarded over the course of evolution, since they ensure that one's genes are passed on.

But what about larger groups like packs, herds and swarms? In the case of a wolf pack, this is a relatively fixed social group in which the individuals all know one another. Depending on the social survival strategy, as a general rule either the females in the group or the males are related to one another for the most part. Here, then, we are clearly dealing with a first-order alliance. A herd, meanwhile, is a collection of animals, sometimes even animals of different species, which assemble but generally do not know one another. They live and migrate together, because they are safer from predators like that, for instance. A herd's behaviour can even give the impression of having been strategically planned. If a herd is threatened, it forms a circle with the adults surrounding the young ones. They all cluster together, and the stronger individuals among them are willing to sacrifice themselves for the weaker.

But what appears so complex and ethically valuable here can

be explained in what are really quite prosaic terms. There is a phrase that is seemingly oxymoronic to describe such behaviour: 'the egotism of the herd'.[53] If the animals failed to behave in this manner, then the young animals would be the first to be slain, and sooner or later the species would die out from a lack of descendants. Conversely, those juvenile animals whose parents were prepared to sacrifice themselves for them will survive. Yet it is precisely within those juvenile animals that the parents' sacrificing genes will also lie dormant, meaning that when the time comes they too will be ready to throw themselves into combat to save their offspring. From the point of view of its function, a swarm, flock or shoal is no different from a herd. The term 'herd' is usually applied to mammals, whereas the former terms tend to relate to insects, birds and fish respectively. In the overwhelming majority of cases, the concept of a first-order alliance would also be apposite in reference to a herd or swarm. This will become clearer, though, once we have discussed what constitutes a second-order alliance.

So, now imagine, if you will, that it's a day later, Monday morning, the start of the working week. You go off to work to earn your crust. At this moment, you're living in a second-order alliance. Such alliances are formed with friends or colleagues with whom you cooperate for a limited period in order to achieve specific goals. Once this has taken place, you go back home to your first-order alliance. A traditional small village community would have functioned in roughly the same way. Everyone had his or her little cottage, but assembled in the fields to work because communal effort made life easier. Biologists also refer to 'fission-fusion societies'. These come together and then disperse. Animals that live in this way are a nightmare for zoos, or rather for other inhabitants, since their social demands are impossible to fulfil. The space at a zoo's disposal is much too small, and there are too few animals to allow a natural lifestyle. I know from my own work with dolphins and great apes that the problems arising from restricted living space can lead to aggression. The only course of action then open to the keepers is to separate the animals, and the upshot of this is often the unhappy one of solitary animals being kept in isolation. Another option is one that we have already seen when discussing the topic of eunuchs – namely, administering female hormones or psychotropic drugs to male animals.

The types of animals that we know live in fission-fusion societies include all the species of great apes,[54] certain whales and dolphins, the African elephant,[55] several predators such as lions[56] and hyenas,[57] but also red deer,[58] giraffes[59] and zebras,[60] as well as bats.[61] There are even suggestions that there might be fission-fusion societies among fish.[62] I'm betting you would not have guessed that the species in question here is the one so widely beloved of aquarists – that is, the guppy.

The one great common denominator of all these animal species is their intensive social life and their ability to cooperate to a certain extent. The more complex their brain, the more complex the form of cooperation is as well. This can extend from common foraging for food and ensuring protection through attaining or consolidating positions in the social hierarchy to exchanging information about resources. In recent years, the number of identified examples of cooperation has grown immeasurably. But one particularly impressed me, so let me take you on a journey straight away to the east coast of Scotland's far north.

This location is home to the last group of bottlenose dolphins resident in the North Sea. You will surely know bottlenose dolphins; 'Flipper' was one, and most of the dolphins in dolphinariums belong to this same species. Bottlenose dolphins occur throughout the world, but even when one has worked with bottlenoses from the Caribbean or the Black Sea, as I have, the North Sea dolphins come as a great surprise. For one thing, they are enormous – at

4 metres (13 feet) long, they are almost twice the length of Black Sea bottlenoses. Their last refuge in the North Sea is a large inlet called the Moray Firth. This body of water is home to around 130 dolphins living in a fission-fusion society.

Researchers from St Andrew's University recognize each and every one of these animals and know who is related to whom. There are two groups living in the firth. One part of the population lives inside the inlet, the other just outside. Now, you might think that the two groups had put some distance between one another in order to each be able to hunt more easily in its territory. But the researchers got a surprise. Certain animals turned out to have good connections to both groups.[63] That is extremely uncommon, since normally both humans and other animals defend their own territory and do all in their power to ensure that, once resources have been secured, they do not have to be shared. Whether this is a fox or a lion defending its hunting ground or we Europeans constructing a 'Fortress Europe' against migrants, from a biological perspective it is all the same mechanism. But the fact remains, if there are deviations from this norm, then this hints at an exciting discovery. If we look at the mapping of the Moray Firth population, we find two independent networks that are connected with one another at various points. These points represent individuals that are active across both groups. It almost looks as though these animals maintain a close relationship not just with family members within their group but also with friends in the other group.

To explain this, one needs to put oneself into the animals' position. Perhaps you have had the chance to dive into the waters of the North Sea, the Baltic Sea, or even an inland lake? If so, you probably experienced extremely limited visibility, since in these locations the water is mostly too murky to see very far. Dolphins don't see much better than you do in such conditions. But they have an acoustic sense that enables them to navigate extraordinarily well. This involves emitting clicking sounds; these are broadband noises whose primary energy resides in their ultrasound frequency, which makes them inaudible to us, and which acts like a laser beam projected ahead of the animal. If this acoustic pulse is reflected by an object, say the swim bladder of a fish, the dolphin can tell the distance and the direction of the 'target' and even what species of fish it is. Unfortunately, this sense only performs really well up to a distance of some 50 to 100 metres (165–330 feet).

In addition, although the sonic beam cannot really be compared to a laser, it is nonetheless extremely concentrated – so concentrated, in fact, that the dolphin risks missing things unless they are positioned immediately in front of it. In other words, a tasty meal might be swimming right alongside the dolphin and yet still remain undetected. It is therefore not all that simple for a dolphin to find food, and is largely a matter of chance. In these circumstances, how handy would it be if someone were to come by and call out: 'Hey, there's big shoal of fish out there right now, on the right side of the mouth of the inlet near the small rock.'

In fact, the researchers think that's pretty much what happens in the case of the Moray Firth dolphins. However, that in turn means that the animals must have developed mechanisms that preclude deception. Animals that heed the call above but nevertheless keep the same kind of discovery to themselves would be at an advantage in the long term, since they would not have to share their food source with others. Over the course of evolution, they would prevail while the animals with a heightened need to share information would die out. In the end, everyone would simply end up working for their own interests and the advantages of cooperative action would be lost.

But if I have a good memory and remember who exhibits what kind of behaviour, then I would reveal the whereabouts of a food source only to those who were prepared to divulge such information to me. In other words, such observations presuppose a high degree of cognitive ability. We must therefore assume that dolphins really do know every individual animal within their network and, like us, are able to distinguish between relatives, friends, distant acquaintances, and those they cannot stand.

This is, without doubt, a striking example, but I wouldn't be at all surprised if similar observations were made in other fission-fusion societies. Until recently, only the great apes were credited with the ability to form such societies. Dogs and wolves, however, which clearly are extremely social, do not inhabit fission-fusion societies. Their stable communities could not tolerate any individuals who transferred their loyalties to other packs as the whim took them. That's a fortunate trait for us humans, for who could bear it if 'his' faithful four-legged companion vanished for a couple of weeks to go in search of a new 'owner'?

Remember our ongoing scenario regarding different orders of alliance? We've already breakfasted in the company of our family, a first-order alliance, on Sunday morning, and then on Monday morning have gone to perform our workaday chores in a second-order alliance. But what constitutes a third-order alliance? Such an alliance forms independently of any routine and presupposes an extremely good, perhaps even lifelong memory. For instance, in a third-order alliance, I call to mind a former colleague who after his undergraduate degree specialized in the study of tardigrades, and give him a call because I'm planning to devote one of the chapters in my book to them. Likewise, our political parties or state apparatuses are alliances of a third or even higher order, since they are based on a stable common foundation that is part of the collective memory of all those involved.

Among the dolphins off Western Australia, this presumably

pans out in roughly the following way: juvenile dolphins establish their own social network from a relatively early age. This network may comprise anything between thirty and fifty animals, or even more.[64] In the particularly intensively studied dolphin population in Shark Bay off Western Australia, there are estimated to be around 2,000 animals, who all know one another more or less well. Each dolphin recruits its own network from this pool. As already noted, males bond for life in very close-knit groups of two or three, and these constitute first-order alliances. When these groups go out looking for a partner, they join together with other groups that they already know well, most probably from childhood, and thereby form a second-order alliance. This insight was based on some twenty years of research in the field. In their third decade of studying this group, the biologists expanded their area of observation in the expectation of discovering that their network model might also predict the behaviour of a much larger group of dolphins. It must therefore have come as a great shock when that turned out not to be the case, and I can only imagine the headaches and the long nights of discussion that must have been caused by the contradictory data. How, for instance, could it be that a larger group, consisting of perhaps three or four groups of three dolphins, disintegrated when a smaller group tagged on? And what possible explanation could there be for the new group turning against elements of the original group? How could a friend so quickly become a foe?

The researchers decided to put a bold hypothesis of theirs to the statistical test by asking themselves whether the behaviour they had witnessed might be explained by positing the existence of a third-order alliance. Prior to this, the only species thought capable of conducting such a complex social life had been us humans. All of a sudden, though, everything made sense, with all the jarring elements becoming plausibly explicable in the light of a third-order alliance. This study is now considered a mile-stone in the investigation of complex social networks.[65] In our democratic politics, people speak in comparable circumstances of party discipline and coalition agreements. For me personally, this example is perhaps the most impressive cognitive achievement of all, since such an observation can only be explained if one presupposes a lifelong memory, an awareness of the self and a conception of others.

If one does this, it opens up a whole new world of possible ways of explaining animal behaviour. The high degree of dynamism observed, but also the great flexibility in forming groups seen among dolphins off Scotland, can be easily explained if we apply models that are used to describe human societies (such as the 'triadic closure model', the 'friend of a friend'[66] hypothesis, or 'Barabási's preferential attachment model'[67]). In the chapter 'Who am I, and who exactly are you?' (see pp. 196–230) we will learn that dolphins possess self-awareness, a name and a conception of others; indeed, that they are even able to use names as points of reference in order to converse with another animal about a third party. If we take this capacity and apply the models mentioned above, then a dialogue between animals might run something like this: 'Hi there, my name's Willy, I'm a mate of Fred's – you know Fred too, right? Can I swim with you?' Reply: 'Oh right, so you're Willy. Yeah, Fred's told us all about you. Sure you can swim with us.'

Crazy, isn't it? And yet this is really how things must play out among animals if the observations mentioned above are to be explained at all plausibly.

Presumably dolphins are far from being an exception. Thus, African elephants can recognize at least 100 animals from the sound of their voices.[68] This even happens when the animals are several kilometres apart. Crows also appear to be just as gifted, for they too visibly react when another individual sounds different from what they were expecting.[69] That's like you suddenly hearing the voice of your good friend A sounding like it's your other friend B. If it weren't important to distinguish between individuals, then this ability would not have developed over the course of evolution. I am very optimistic that we will in time discover many more complex social communities among animals.

19. The invention of morality

My mother-in-law was appalled: 'What, you allow your children to be experimented on!?' Together with the discharge documents following the birth of our twins, we received a consent form from the University of Erfurt stating that we were in principle willing to take part in experiments researching into young children.[70] To be honest, I was actually delighted at this. In my world, such experiments are a definite plus; I assumed that my children would feel the same, and indeed they haven't complained so far. When the day for the first set of tests came, it turned out to be about colour recognition.

The parents are given some brief guidelines on how they should conduct themselves so as not to distract or influence their children. At the end of the experiment, the female student researcher was surprised when I asked her whether this was a 'false belief' test. Of course, such experiments aren't about determining whether Christianity is a false belief from a Muslim perspective, or vice versa. Instead, they are designed to find out whether the test subject wrongly believes an assumption of theirs to be correct, and whether he or she can conceive that others might be proceeding from a false assumption. That is extremely important, for example, in trying to anticipate the behaviour of other social partners. If their behaviour is based on a false assumption, then any plans I make might be well wide of the mark even if my own assumptions are correct. In a complex social life, one is constantly called upon to make these kinds of decisions, but in actual fact it all comes down to equitable distribution.

Let's go back to our dolphins in Scotland. For instance, a dolphin thinks: 'No, I'm not letting him know where there's a big shoal of fish right now; after all, that's only right and proper as he's never ever tipped me the wink.' However, it might be the case that the first dolphin only thinks that he's correct, for in fact the second dolphin has given others information on plenty of occasions. You see how difficult it is to reach the right decision in a social community that is based on reciprocity. But is it even possible to test such things in an experiment? The first false belief experiment my kids took part in didn't have to do with colours as such, or fish for that matter, but rather pirates.

The children who were taking part were told the following story: A long, long time ago, when bold pirates still roamed the Caribbean, an exhausted pirate sat down at his table and put a sandwich on it. It suddenly occurred to him that he was thirsty too, so he left the table to get himself something to drink. Now, it's often very windy in the Caribbean, and while the pirate was gone, his sandwich was blown clean off the table. Soon after, another pirate came along and put his sandwich down on the table. He also felt thirsty and went off to get himself a drink. No sooner had he gone than the first pirate reappeared and sat down at the table to enjoy 'his' sandwich in comfort. At this juncture, the children were asked, in passing: Which sandwich will the pirate eat?

The situation is crystal clear to us adults. The first pirate comes back, and has no idea that his sandwich has been blown off the table and that the one sitting on the table in front of him actually belongs to the second pirate. Our response to the question would be that he eats the one that's on the table. But to children aged three, the world looks quite different, and so their reply is that the pirate would pick up his sandwich off the ground. Anyhow, the story now continues, but they're told that the pirate takes the sandwich that's on the table. Whereupon the child being tested is asked: Why did the pirate take the sandwich on the table, then? In response to this, the children come up with various stories such as: Maybe the sandwich on the ground had got dirty and the pirate didn't want it any more. The key to this reasoning is an enigma familiar to all parents. The world inhabited by our small children is a paradise. In this world, Father Christmas is able to visit all children at the same time, no matter where they live.

By the time children reach the age of five, the picture has already changed. They are well aware that the pirate cannot know that the sandwich on the table isn't actually his. But if they are then asked if it was right of him to eat the sandwich on the table, or whether he was bad to take it and ought to be punished, they all reply: Yes, he needs to be punished because it wasn't his sandwich. In their world, the pirate hasn't behaved properly. Despite the fact that they can understand the situation, they are still unable to get a moral handle on it or to recognize criminal responsibility. Yet what is remarkable is that if, according to their conception of the world, children see an injustice, they take action against it. This remains the case even when doing so entails a degree of effort on their part and they personally derive no benefit from it.[71] In other words, they stand up not only for themselves but also for others. Could it be that we have an inbuilt sense of justice?

For us adults, the situation is perfectly clear – the pirate is inno-cent. He might well end up swinging from the gallows because he robbed others in order to be able to afford his sandwich, but he oughtn't to be punished for eating the wrong sandwich. I recommend to anyone who is interested in looking at this in greater detail that they go online and watch the TED Talk delivered by Rebecca Saxe, the researcher who devised this particular experiment.[72]

Someone who, while cleaning his friend's gun, accidentally shoots him dead is every bit as innocent as our pirate. Lawyers the world over would concur that this does not constitute murder, but instead was an accident. A part of the brain called the rTPJ area – named after its position, the right temporoparietal junction – is responsible for making such decisions. It is only as big as a fingertip and is sited in our brains just above the ear. It is active whenever we are considering what other people think. For as long as this part of our brain remains undeveloped, we are incapable of evaluating matters in the way described. Children under the age of seven are unable to make this kind of moral judgement call.

Through the use of strong magnetic fields, it is possible to temporarily paralyse certain areas of the brain. A very good friend of mine, Bas Kast, once subjected himself to just such a procedure in the course of researching his book 'How Your Stomach Helps Your Head Think'. His interest was in creativity, and indeed his brain did turn out to be more creative when certain areas were switched off. But what exactly happens when the rTPJ area is blocked? Yes, your guess is correct – we are no longer capable of arriving at moral decisions, our moral organ is silenced and we temporarily lose the ability to empathize with the reality of others.

Before we come to look at morality in the animal kingdom, I wanted to make clear that the foundation of our morality is not supplied by Kant, say, or any other of the great philosophers. Rather, the basis of our morality resides in our brain's capacity to make moral judgements. This capacity develops in the first years of life, and as a general rule children of that age do not know much about Kant. As such, it is a perfectly legitimate exercise to enquire after this ability in other animals also. But what exactly are we looking for here, what is the common denominator of all moral questions?

For most people, the answer lies in religion. But what happens if we apply the insights we have just gained concerning false belief to religion? A Christian might maintain that a Muslim has a false belief. He might regard Muhammad as a charlatan and take the view that all Muslims are wrong to see him as a Prophet of God. For a Muslim, though, it may well be the exact opposite; according to him, all Christians would be heathens for refusing to recognize God's last prophet. Both sides are of the opinion that the other has a false belief. The only question is: what form does this moral judgement against the other side take? Is the other side to be punished for its false belief? Seen from a non-religious neutral standpoint, of course not, since both sides cannot help holding their supposedly false beliefs.

Now comes the point that I'm truly interested in: if one or other side starts killing adherents of the other faith, or if Turkey sends Dutch cattle back to the Netherlands because its president is denied the right to hold an election rally for Turkish émigrés there,[73] then I have two possible ways of viewing the situation. I can work either from the assumption that the people concerned are intellectually on the level of a six-year-old or that their morality is so underdeveloped that it is not inclined to resolve the situation through social consensus. So, if we are looking for moral behaviour then on no account should we navigate by a morality that we have developed within the context of our culture; instead, we need to search for primal mechanisms operating within simple social systems.

Once again, I come back to my dolphins in Scotland. In their social community, it is imperative that no animal should profit at the expense of the rest by disclosing no information itself while benefitting from that divulged by others. After many years of discussion, behavioural biologists now think they have identified fairness – also referred to as 'inequity aversion' – as a fundamental social mechanism. Should a dolphin fail to disclose information, then it is behaving in an unfair way towards the other members of its social community. Likewise, if a person of faith deprives another believer of his or her life or land, then they too are acting unfairly.

Unfortunately, observing and documenting fair behaviour in the wild is very complicated. On the other hand, a number of experiments have been devised to test fairness. All that is required is to treat two animals unfairly.

Frans de Waal was the first researcher to attempt such an experiment. He had his students feed two capuchin monkeys. Of course, they only received their food in return for performing some small task. In this case, it involved passing a small stone through the bars of their cage. For each pebble, they received a piece of cucumber. But after a certain time, one of the two monkeys was rewarded with a grape instead. Monkeys have much the same attitude to cucumbers and grapes as we do, and so the monkey with the cucumber soon began to feel it was being treated very unjustly. Consequently, in sheer exasperation it tossed the next piece of cucumber it was given back in a high arc to the researcher. If you look at the video,[74] you almost feel that the little animal is on the point of tearing its cage apart in fury and going for the researcher's throat. Frans de Waal was therefore able to demonstrate that capuchin monkeys can perceive injustice.[75]

Fairness is also a major topic as regards bringing up sibling children. If, for example, I want to dress my two sons in different coloured socks, it only works if each of them is allowed to wear one sock of either colour. The so-called inequality aversion is deeply ingrained in us, and twins aged three or four are a prime example. As a parent, it's sometimes helpful to remind oneself of this when you're at the end of your tether.

Following the great success of the capuchin monkey experiment, chimpanzees were also successfully tested.[76] However, the research rather hit the buffers when it failed to elicit the same good result from orangutans.[77] As you know, orangutans are one of the four anthropoid ape species (humans, chimpanzees, gorillas and orangutans), and we customarily set great store by their abilities. The researchers similarly drew a blank with squirrel monkeys.[78] That too came as something of a surprise since, like capuchins, squirrel monkeys belong to the Cebidae family, and Frans de Waal had had great success in experiments with the former. But even though the squirrel monkey is the species famously pictured as the pet ('Mr Nilsson') sitting on the shoulders of Pippi Longstocking in the eponymous series of children's books by Swedish author Astrid Lindgren, like orangutans they are not very social in reality and live mostly alone or in mother-offspring pairs. As loners, orangutan males are in a class of their own – in the evenings, they call in the direction in which they wish to travel the next day so that all other males along the path of their intended route will clear out of the way.[79] They use their fat cheek pouches like a megaphone to project the sound.

Lest you get the wrong impression about orangutans, though, there are plenty of examples of culture[80] and tool usage among this species as well. Besides, calling out a day in advance is a clear sign of planned action. Orangutans are therefore by no means the dunces among the great apes; they are just not very sociable creatures and hence have not needed to develop any mechanisms for equitable cohabitation.

By contrast, macaques lead a very sociable existence and – *quelle surprise!* – they were successfully tested for inequity aversion.[81] This only serves to corroborate the suspicion that inequity aversion had less to do with how highly evolved an animal was and far more with the nature of its lifestyle – that is, social or non-social. That is remarkable in so far as this cognitive achievement presupposes various intellectual accomplishments. Thus, for example, I have to be aware of the existence of someone other than myself. We will deal with this topic at greater length in the chapter 'Who am I, and who exactly are you?' (see pp. 196–230), but for the vast majority of animals, other animals are nothing more than an interactive component of their environment. They are unable to empathize with them or to imagine that these other animals are individuals capable of independent action. In addition, in order to be able to perceive injustice, one needs to have an understanding of quantities or of set theory. One also requires a memory in order to know that one has just cooperated with someone else who is now demanding his rightful quid pro quo. Certain other items could no doubt be added to this list.

Despite the high degree of cognitive ability demanded, many researchers still felt emboldened to try testing other mammals and birds, such as dogs[82] and ravens.[83] All very successfully. Even rats were tested successfully.[84] I am therefore convinced that we will discover the foundations of morality in many, many more animal species. Presuming, that is, that they lead a social life in which participants need to rely upon cooperation and fairness.

Granted, fairness is far from being the same thing as morality, but it is a first step all the same. If I am in a position to draw comparisons within my social community, then I am in a position to judge whether something is just or not. Bonobos, for instance, protest vociferously if they are not treated in strict accordance with the acknowledged social rules of the communities within which they live.[85] Isn't this the ultimate basis of our entire moral system? Nevertheless, the animal is not thinking here in moral terms; first and foremost, it is thinking of itself.

In this respect, the behaviour of the capuchin monkey that is given the grapes is far more interesting. Is there such a thing as solidarity? It certainly exists in human society – children develop this behaviour when they are somewhere between six and eight years of age.[86] In the experiment with capuchins cited above, in order to act in solidarity with the other monkey, the capuchin that had received the grapes would have had to relinquish them. Hard though it is to credit, precisely such behaviour was actually observed among chimpanzees.[87]

The question that now arises is: how does one behave towards someone who has behaved unfairly? Of necessity, a social community only functions properly when antisocial behaviour in any form is associated with disadvantages. Among our ground squirrels, we saw that cheats were simply ignored and received no reward for keeping a lookout. Measured against the way in which we humans deal with cheats and criminals, this is a lenient punishment. Up until relatively recently, a person who had stolen something would have had their hand cut off. Nowadays, society punishes thieves not by depriving them of their hands, but by depriving them of their freedom by locking them up in prison. We humans thus live in a society in which antisocial behaviour is punished by force. Incidentally, that's not so very different among capuchin monkeys, despite the fact that they haven't yet invented the severing of hands or prison.[88] Capuchins punish antisocial or uncooperative behaviour by displaying aggression. However, researchers no longer refer to aggressive behaviour in this context, but to punishment. This is in order to make it clear that transgressions against the rules of cooperation have taken place. Punishment, therefore, is associated with an effort that doesn't directly pay off. If a dog snaps at another in order to grab for itself the food that its fellow canine is eating, then it is gaining for itself an immediate advantage in nutritional terms. Punishments, however, tend not to have an immediate impact on canine behaviour. The hope is that the act of disciplining will bring about improved social behaviour in the future. This is an important point with far-reaching social consequences, since it presupposes that animals can conceive of what lies in the future.

We humans go a step further, for, unlike all the animal species we have looked at thus far, we are the only one in which individuals who are not directly affected by a case of transgressive behaviour are the ones who enact punishment for it.[89] Just picture this scenario, if you will. Someone comes along who has nothing whatsoever to do with the whole situation and weighs in against an individual who has violated certain social norms of behaviour. One is tempted to say 'Why don't you mind your own business?' But behind their action there lies an incredible achievement.

We're dealing here with an individual who has nothing to do with the conflict, and yet who is prepared to take a risk in order to ensure that the rules of society are observed. This person does not gain the slightest measure of instant benefit from their action, and even in future it will be the whole of society that profits rather than the person who enacts the punishment. Research on this topic really set me thinking, and has made me wonder whether such studies might begin to fathom the pronounced human inclination towards the use of force and brutality. You may perhaps be familiar with the 2015 film *The Stanford Prison Experiment*. It is based on a true incident from 1971. The psychology professor Philip Zimbardo from Stanford University in California conducted an experiment in which he asked a group of young people to play the role of prison inmates, and another group their warders. After just a few days, the level of violence began to escalate between the two groups and the experiment had to be terminated.[90]

But although there is much to be said for the thesis that violence has made us humans what we are, we now know that individuals, be they animals or humans, or even entire societies, can be more easily controlled through positive reinforcement than by the use of force. The technique of positive reinforcement is used by practically every professional animal trainer, and is the basis of most modern pedagogical theory. I felt obliged to say this lest I found myself accused of making excuses for the use of violence.

From the standpoint of behavioural biology, though, we are only beginning to scratch the surface of this question. There is no shortage of researchers who start from the premise that norms, and hence normative behaviour, also exist within animal social communities.[91] In such circumstances it would be an irrelevance who was being punished by whom and under what conditions. For example, I have heard anecdotes concerning a number of dolphins swimming head-on, as a group, towards an individual. Now, if I tell you that face-on swimming is an extremely impolite and aggressive posture for dolphins to adopt, you can well imagine how an individual dolphin must feel on seeing a whole pod swimming directly towards it. To anyone looking on, it seems very much as though an individual animal is being punished by the community. Indeed, my former colleague at Dolphin Reef in Israel, Elke Bojanowski, has discovered that mother dolphins use a special disciplining whistle to 'scold' their unruly offspring.[92]

Many researchers can cite numerous such examples or anec-dotes, yet it is very rare for these individual observations to be published. As a result, I was especially delighted by the following example from Leipzig Zoo.[93] The article recounted the following observation: Three adult chimpanzees X, Y and Z along with three juveniles 1, 2 and 3 were sitting in their enclosure. The young chimps 1, 2 and 3 were playing with Z's blanket. The adult chimpanzee X proceeded to chase away the youngster 1, grabbed the blanket and disappeared up the nearest tree. The adult chimp Y now picked up the blanket that X had just been sitting on and chucked it to chimp Z. Soon afterwards, the three young chimpanzees started playing with a blanket again, only this time it was X's blanket, which chimp Z had given them to play with, just as he had allowed them to play with his own blanket beforehand. Evidently, chimpanzee X was an 'antisocial egomaniac'. He already had a blanket of his own, but felt compelled to demonstrate his dominant position over the young chimps by stealing the blanket they were playing with.

What's interesting is the behaviour of chimp Y, who took X's original blanket and gave it to Z. It is really hard not to impute to chimp Y an honourable attitude and a sympathetic understanding of the situation. The authors of the article omitted to give a final explanation of the situation, but their observations encourage us to take a careful look at these kinds of interactions in zoos or in the wild.

20. Death cults and war

The air was heavy and the early morning mist rising from the grass smelled of earth, life and death. Beneath the dense canopy of leaves, the rising sun produced a kind of twilight through which a patrol of combat-ready males made their way in disciplined single-file formation through the undergrowth. The awakening jungle was full of noises, but the group moved quietly and warily. Their eyes darted left and right, scanning their surroundings for potential quarry. They knew they were in enemy territory and that it was only a matter of time before the next deadly skirmish broke out.

It was a beautiful morning. The youngsters were playing boister-ously and had failed to notice that they had strayed much too far to the south. They knew the area well and had their favourite play spots, but latterly there had been a spate of attacks. The two playmates had already lost several friends hereabouts, but that had slipped their minds for the present, caught up as they were in their enjoyable roistering. One of the youngsters had just picked up a suitable stick to use as a weapon and was looking forward to playfully whacking his pal with it. Suddenly, without the slightest warning, something large and dark-coloured burst out of the dense undergrowth. The youngster saw his friend surrounded by five powerful adults, and ran off screaming before he had a chance to see them rain blow after vicious blow upon his friend's small and unprotected body.

These guerrilla tactics were employed not by humans to terrorize others but by the clan of Ngogo chimpanzees in Kibale National Park in Uganda to expand their territory. Yet this was not normal behaviour; one might even go so far as to say that such warfare is abnormal, since it runs counter to a mechanism that is widespread in nature, namely the so-called endowment effect. Without this effect, our world would be simply chaotic, an apocalyptic place where no one would want to live. This fundamental biological mechanism was discovered not by behavioural biologists but by economists.[94] These researchers were surprised to find that people valued their own possessions far more highly than they did the comparable possessions of others. The upshot of this in the real world is that a person is prepared to invest much more in retaining his or her own property, or to fight far harder to keep it, than another person is in trying to thwart them. Picture yourself as a butterfly flying towards a flower. Your tranquil world is soon upset, however, as another butterfly appears and tries to muscle in on your flower. You are genetically programmed to fight for your possessions, and can keep on doing so ten times longer than your attacker. The attacker is aware of this state of affairs and respects things that are already spoken for, so in the normal course of events would not attack in the first place. In other words, respect for others' property is genetically deeply ingrained. In contrast to most animal species, which only break out of this peace-promoting programme loop in the direst emergency, we humans – but also other cognitively highly developed animals like chimpanzees – have the potential to override this mechanism.

Maybe you'll think me somewhat perverse, but for me this kind of mechanism represents the beauty of nature every bit as much as a flower, a baby tiger or a glorious sunset. Of course, I didn't just pluck the example of the butterfly out of thin air.[95] In fairness to behavioural biologists, I should perhaps mention that the publication about territorial defence in butterflies appeared two years before the economic study, though at that stage the term 'endowment effect' had still not been coined.

But to return to our Ngogo chimpanzees in the Kibale National Park in Uganda, you can find a video sequence of similar behaviour on YouTube.[96] There you can also see the patrol routes of the chimps and those of their victims in the period from 1999 to 2008. In that time, they expanded their territory by 22 per cent and expelled the former occupants. The takeover was finally sealed in 2009, when females and juveniles of the Ngogo chimpanzee group were sighted for the first time in the new territory.

By adopting the strategy of advancing stealthily into foreign territory and carrying out targeted killing of young chimpanzees, the Ngogo chimps succeeded in displacing the original inhabitants. This example[97] is the first documented case of a deliberate act of war outside human society. Even so, it is far from being an isolated case. In the 1970s the Kasekela chimpanzees in Gombe National Park in Tanzania also conquered the neighbouring territory of the Kahama chimpanzees.[98]

Leaving aside any moral considerations, the strategic dimension of this action is quite incredible. The animals who went on patrol were acting quite differently from the mode of behaviour they displayed in their everyday activities. Normally, chimpanzees move as a disparate group and aren't exactly quiet about it. Here, though, they kept quiet and advanced in single file, and did not allow themselves to be distracted by food sources or by searching for female chimps. Their clear purpose was to carry out targeted, murderous attacks, and the duration of the latter gives us much food for thought.

In the video mentioned above, the Yale professor David Watts presents the results of his study of the chimpanzee war. If you've watched the video, perhaps you too were struck by the fact that he is being interviewed against the background of a military cemetery. War and the cult of death, as well as the idea of a warning to succeeding generations, are all closely related topics, but is there also such a thing as a death cult in the animal kingdom?

The *Tarzan* movies helped popularize the myth of the elephants' graveyard, a place where older elephants go to die, of their own volition. In reality, however, every year the fate of thousands of elephants is sealed by human agency – in particular by the activities of poachers. Our appetite for pretty carvings and keys for expensive pianos has driven this incredibly highly developed animal to the brink of extinction. According to an article that appeared in the journal *Nature*, surviving elephants have been robbed of their culture by the selective loss of the larger, more experienced and dominant animals and will in all likelihood never live again in the same way as they did before the great genocide that has been perpetrated against their species.[99] The article is at pains to point out that elephants can only be successfully protected if one ensures that their culture also remains intact. Just like us humans, elephants are scarcely able to survive without their culture, as they are deprived of important modes of behaviour and knowledge.

Some genuine elephants' graveyards *have* been found, in the form of 'cemeteries' of long-extinct woolly mammoths. Perhaps this discovery was what formed the real germ of the myth. Yet in all probability, even these woolly mammoths did not assemble at these 'cemeteries' in order to die. At some stage, the teeth of older elephants wear down and they take to eating softer vegetation. They find these softer plants in damp areas, which for the most part are not very extensive in Africa, which explains the unusually high number of skeletons found at certain locations.

Even though these 'graveyards' appear to be instances of chance accumulations of mortal remains, it is nonetheless true that animals foster a cult of the dead. For example, one can observe how they cover their dead with foliage and branches and, in this way, inter them in the truest sense of the word. Elephants also bide awhile when they encounter skeletons of animals that they once knew.[100] This means they must comprehend that the pile of bones in front of them is a dead elephant, and if one also takes their other cognitive abilities into account, then one is surely justified in concluding that these animals have a conception of death and know exactly whose remains they are standing in front of. One research project on elephants produced a quite remarkable, but very well documented, anecdote. It concerned the death of a dominant female. The research data suggests not only that she was mourned for several days by the animals of her own clan, but that other animals came from far and wide to 'pay their respects'.[101]

It is also worth mentioning in this context that elephants are reported to weep at moments like this.[102] Another interesting observation is that communication among the animals after the death of a close relative changes quite radically. All in all, it is hard to interpret these signs as anything other than clear evidence that elephants have a conscious awareness of grief, even over a protracted period.

But what actions, exactly, constitute mourning and a cult of the dead? Is a dog that lies down at its master's grave grieving? Naturally, as a sympathetic being I register its pain, and I wouldn't bridle at anyone who interpreted this as grief. Yet from the perspective of behavioural biology, things are not that simple. The dog may at that moment have lost its pack leader and possibly even its entire social network. Inevitably, this situation causes it suffering, but is it really mourning? Is it grief that makes it refuse to eat or is it rather the loss of its social sphere? Of course, one could also imagine a human being in a similar situation, and who can say for sure whether that individual is mourning their nearest and dearest, or is in despair at the prospect of being alone following their loss? However, there are two essential indicators of genuine grief.

First: what treatment is accorded the dead body? There is a wide spectrum of potential behaviours in such a situation. In nature, a cadaver is a source of energy. Cannibalism would therefore be a perfectly sensible strategy. For many animals that eat meat, this would also be a normal reaction. To do otherwise would just be a waste. It's a different picture for herbivores, which ignore a dead body as far as possible. A cadaver can also provoke aversion, for instance when it is in an advanced state of decay. Ultimately, the corpse can also remain part of the social community. In this regard it doesn't matter whether, as with elephants, it is in the form of a skeleton or, as in human cremations, a pile of ashes in an urn.

Second: in order to really be able to judge whether an animal is in mourning, one needs to observe its customary social life. An animal that suddenly undergoes a change in its circumstances (such as the death of a social partner) but does not mourn will still display behavioural abnormalities that manifest themselves in its conduct in everyday situations, whereas an animal that is genuinely mourning will behave relatively normally within its social sphere and only show abnormalities around the corpse. With us humans, things are somewhat different. In the case of bereavement, we avail ourselves of the luxury of being able to forego social interactions, since we do not have to worry about our immediate survival. We have credit cards and can get everything we need in a supermarket. Animals, on the other hand, have to get on with their normal lives without pause. They have to keep watch for predators, or find a sheltered place for the night or the next watering hole. But despite all this, some are still capable of experiencing grief, and the task of scientific research is to document precisely those cases.

As much as fifty years ago, chimpanzees were frequently seen to behave in a very remarkable way towards deceased members of their species.[103] Especially in the hours immediately after death, for example, the body of a dead juvenile chimpanzee would be given a thorough examination by members of its group, and particularly by its mother.[104] Some mothers carried their dead and mummified babies around with them, in some cases for weeks or even months on end, thus taking upon themselves a senselessly high burden of effort. In such circumstances, there could be no doubt that the animal was genuinely grieving.[105] But in the main these were isolated instances, meaning that one could certainly speak in terms of mourning, but not of a death cult. In order for such practices to amount to a cult, they need to be fully integrated into the broad culture of the species in question. Yet that is precisely what was observed some years ago.[106] In a chimpanzee population in Guinea, no reactions of aversion or cannibalism were noted, but instead it appeared to be part of the chimps' culture to artificially mummify deceased young animals and carry them around. The cadavers of mammals only mummify under special conditions, and so either chance comes into play here or the animals are somehow actively promoting the mummification process. They do this by constantly shooing away flies and plucking insect larvae from the corpse. This exact behaviour was observed as a common cultural feature across the whole population and so, as with the elephants, one is forced to conclude that a genuine cult of the dead is in operation here.[107] The researchers ended their report on a decidedly sombre note.[108]

Chimpanzees are not the only animals that carry their dead around with them. Similar observations have been made of at least two different species of dolphin,[109 & 110] while a search on YouTube using the term 'dolphin carries dead calf' produces many hits.

After having looked at morality, conflict and death cults, let us now turn our attention to our financial system. I can promise in advance that this will give you not only plenty to smile about but plenty to ponder, too.

21. The brokers

Not long ago, evidence was published for the survival of an order of mammals once thought extinct.[111] The group in question was the highly social rhinogrades, which were first described by Professor Harald Stümpke (Zoological Institute of the University of Karlsruhe) in his 1961 monograph 'Form and Life of the Rhinogrades'.[112] Although these creatures were of considerable scientific interest, the public knew virtually nothing about this order of mammals, which was endemic to the South Pacific. Their most striking characteristic is their long, snout-like nose, which they use to propel themselves along. This earned them the popular name of 'snouters' in the English-speaking world. I don't intend to go into their unconventional physique and their bizarre-sounding method of locomotion here in any greater depth, since their social life is yet more interesting.

As a result of their isolated lifestyle, over the course of evolution various species of rhinograde have come into being. These species could be said to correspond to the functions that we humans carry out in our society. Apart from a very few exceptions, all the different species of rhinograde live in a social network in which a variety of tasks and resources are shared. This is a hitherto quite unique arrangement within the animal kingdom and only works through their use of a common accounting unit.

Of course, the rhinogrades don't *really* have money, but – like the first aboriginal humans to inhabit the South Pacific region – they do use rare seashells as an equivalent. There appears to be one particular species that does nothing except lend these rare seashells. After precisely one phase of the moon, these shells have to be returned and, in most cases, a few more shells are brought back than were originally lent out. These animals therefore display an understanding of set theory and added value.

Because of this special adaptation, this species has naturally been the subject of considerable interest. In contrast to most other species of rhinograde, very few individuals belong to this particular species – commonly and jokingly referred to as 'brokers' – and they lead an extremely precarious lifestyle. Presumably in order to better protect their 'wealth', they occupy rare cliff habitats with a splendid view, which enable them to see changes in the weather or the approach of enemies well in advance. In these circumstances, however, their apparently clever behaviour in lending seashells can lead to their undoing. Space on their much sought-after clifftop locations is very limited, and the animals hoard their shells there. And because the heaps of shells keep growing larger without them having to do a thing, since the other species invariably pay them 'interest' in the form of a few extra shells, the available room on the cliffs is steadily shrinking. This can result surprisingly frequently in one of the animals slipping on its shell pile and plummeting to its death over the cliff edge. As a result of its extraordinary lifestyle and the threat facing it, this species has been classified as critically endangered by the IUCN,[113] the leading international animal conservation organization, and placed on its 'Red List'. However, this listing made it possible to fund appropriate conservation measures, with safety nets being installed beneath the precipices.

Now, I hope you won't be too irked when I tell you that this whole thing is an elaborate hoax, a pure confabulation designed to point up the crazy behaviour of a species of great ape that we know all too well. The idea of the rhinogrades can be traced back to an old biologist's prank dreamed up by Professor Gerolf Steiner of the Karlsruhe Technical University on the basis of a verse by the absurdist poet Christian Morgenstern. Since the appearance of his 1961 monograph, which he published under the pseudonym of Professor Harald Stümpke, his work has been picked up several times in serious scientific literature, to the point where it now enjoys something of a cult status. But because Steiner's original treatise barely dealt with rhinogrades' complex social life, I thought I would spin the yarn a bit further. But don't laugh too soon, as the real circumstances in the animal kingdom aren't so very far from the fantasy. Thus, in the chapter 'Against reason' (see pp. 231–41), we will learn that the safety net provided for our banks, not to mention the root cause of the global financial crisis of 2007–8, can be identified in patterns of behaviour that are some 30 million years old, and that the added value of an abstract accounting unit is also capable of being devised by animals (see the chapter 'Mathematics'; pp. 191–2).

V

On thought

As a child, I was enthralled by Ancient Greek history, and the subjects that particularly captured my imagination were Atlantis, Plato and the oracle at Delphi. In later life, when I found myself actually standing outside the entrance to the oracle at Delphi and enquired where I might see the inscription '*Gnothi seauton*', I received the following disappointing reply. 'On a column at the Temple of Apollo, if it was still standing.' If only I'd taken the time to read a travel guide, I'd have been spared this embarrassment, for the inscription *Gnothi seauton* ('Know yourself') was long since known to have disappeared and has only come down to us through the reports of contemporaries. But it was precisely this that motivated me to expand my consciousness, experiment with certain drugs and take part in meditation. I was on a quest to find myself, but the spur to do so came from a past civilization that was 2,500 years old. At that time, we humans began to discover and comprehend the world in a systematic way and to preserve for posterity the knowledge we acquired. This inscription is the cornerstone of Western society, in which each and every individual counts for something and is protected by human rights legislation.

In contrast to the themes we have treated thus far – sex, social life and culture, which can only exist within a community with other species members – let us now turn our attention to the question of the individual.

22. *If you think that you're thinking, then you're only thinking you're thinking*

As a rather self-possessed schoolboy, one Sunday morning I found myself being woken from sleep with these words ringing in my ears. The radio was playing, my parents were in good spirits, the picnic hamper was packed, and the Trabant was ready for the off with a full tank of petrol. Although I didn't understand the words of this popular song by Juliane Werding – since ultimately it was about a bloke who pulls lots of girls – I still found the wordplay amusing. That's hardly surprising when you consider that it's not uncommon for people at the age I was then to mull over their own thought processes (see part V 'On Thought'; pp. 169–268). Don't worry, though, I'm not about to sit you down in the Trabant with my family and take you on an excursion into the countryside; instead, let me take you on an airliner bound for Majorca, where we're about to pay a visit to the beach promenade at El Arenal. Every few metres or so along the seafront, there is a pitch where, with the utmost perseverance, a guy is conducting intelligence tests. It's the best place imaginable for this, as the test subjects are relaxed, and anyone who takes the test once never comes back for more. As a result, a large number of test subjects can be processed in a single, small location. What's being tested here is so-called object permanence. To do this, a small object is placed under one of three inverted cups, and as the cups are quickly shuffled by the experimenter, the observer has to keep an eye on them and remember which cup the object has been placed under; finally, he has to pick the cup beneath which the object lies. If he gets it right, he wins 50 euros and he passes the object permanence test.

In the normal run of things, from the age of two onwards we humans have no difficulty doing this, but on the promenade at El Arenal nobody ever passes the test, and I suspect that what's really being tested there is people's tolerance of frustration.

Perhaps the criminal shenanigans of a few locals in El Arenal don't have anything much to do with scientific experiments, but the abilities that we are confident we possess, and why people keep allowing themselves to be taken in by this scam, have to do with an essential property of our brain. We are perfectly capable of retaining thoughts, images or objects in our brains, even when they are no longer there in reality. Without object permanence, only the simplest of thoughts would be possible.

Mental images

If we really want to try to picture how animals think, we must first get a clear idea of what thinking actually entails. The renowned French philosopher Joëlle Proust has put forward a very simple but practicable model, consisting of just four stages.[1]

In **Stage A** (stimulus–response), animals are capable of receiving stimuli and reacting to them. For instance, if you poke a snail, it will retract its telescope eyes and possibly also its entire body into its shell. This is a reflex response that the snail can do nothing to influence. Though frequent stimulation can result in a weakening of the response and an adaptation to it, the core reaction remains; there is an external stimulus that leads directly and unalterably to a consequent reaction. However simple this might sound, many things are in fact governed via this mechanism, including foraging for food, looking for a sexual partner or protecting oneself from harmful environmental influences.

In **Stage B** (protorepresentation), a mental image of the environment is formed for the first time. It is no longer a question of a stimulus followed instantly by a response, but rather the stimulus triggers a chain of mental processing. The object that has caused the reaction is visualized in the mind. Scientists therefore also refer to this action as 'protorepresentation'. The choice of term is unfortunate, for surely something is either mentally represented or it isn't. How can something be represented in advance? The term is intended to underline the fact that the representation is still dependent upon the external stimulus and hence not yet a fully intellectual construct. In principle, protorepresentation should only facilitate a verification through another sensory organ. If the mental presentation of both sensory impressions is in agreement, the probability of a false interpretation is greatly reduced. Glancing in the direction of a rustling sound in the undergrowth reveals to us either a blackbird or a panther. Now, it would of course be stupid to flee from a blackbird. What's in play here, therefore, is a kind of control mechanism. Since this representation is captured mentally, the object that has caused the stimulus possesses a certain conceptual permanence, and this leads researchers to talk for the first time in terms of the capacity of object permanence. This has a further advantage, since where a predator is concerned it is a pretty practical matter. Just imagine that your prey suddenly disappears behind a tree. In the absence of object permanence, the predator would in that same instant forget the quarry it was actually chasing. However, thanks to this retained mental image the predator knows that its prey cannot really have disappeared, and so makes a move to flush it out. For this reason, many animals do not do at all badly at their version of the 'shell game'.

In **Stage C** (category formulation), things get a little more complicated. Here, animals must be capable of formulating categories. Perhaps you will recall our bowerbirds. If a bowerbird collects objects that are coloured blue, then it can only do so because it has created a category for blue. Without any external stimulus, it must think about collecting blue objects. This is an enormous achievement in terms of object permanence, since it must have kept the category for blue objects filed away in its mind the whole time. If you as a person think about blue objects, then in that moment you are precisely replicating the bowerbird's thought process. In the model, and probably also in reality, there is no difference. When animals such as coal tits can tell the difference between an alarm call and a mating call, then these calls too have an object permanence and are organized into categories. The same applies to pigeons, which can distinguish between images of female and male humans[3] or are capable of recognizing two things – for instance, two one-cent pieces – as being of the same kind and picking out two similar grapes.[3] There are even some animals that can distinguish between a Monet and a Picasso. That's not so difficult, you might think, but you'll be amazed to learn how small the animal in question is (it's the honey bee!).[4] Categories are an extremely useful thing, since they reduce the mental computational element and help animals come to swift decisions.

Stage D is full mental representation. The mental images in this phase no longer have any connection to the original stimulus. This opens up the potential for completely free play with the mental representations. At this stage, I am capable of so-called metacognition. In other words, I can reflect on my own thought and, for instance, revise decisions that I have come to or ascertain that I do not understand something and require more information. This is a great thing when I cannot simply reach objectives directly but need a strategy to get closer to them. Anyone who has been surprised by a sudden cry of 'Check!' from the person they're playing chess against will know what I mean, while metacognition will enable them to reflect on what they did wrong.

Using this stage model, it is possible to describe every mode of thought on our planet and even to test them through experiment. For example, object permanence is tested through the use of the 'shell game' that we have already mentioned. Frequently in such experiments, researchers use the classification system devised by the famous Swiss developmental psychologist Jean Piaget, who specialized in child development. So as not to bore you, we will only concentrate here on stages 4, 5 and 6.[5]

At Stage 4 in Piaget's classification, animals are able to locate a morsel of food that has been placed under one of three cups. A researcher picks up the food, lifts up one of the cups, places the food underneath it, and then leaves the animal to decide which cup it will search beneath. That sounds simple, but the animal requires at least the capacity of protorepresentation, for without a mental image the stimulus disappears as soon as the food is placed beneath the cup. An animal without this capacity would have to look beneath each cup. Human children are capable of proto-representation by the age of around one year.

How highly developed the capacity for protorepresentation is becomes apparent when the cups are switched around, like the scammers do at El Arenal. If the animal is able to follow this switching process, then it has reached Piaget's Stage 5. Human children reach this stage between the ages of twelve and eighteen months. Animals that tested successfully at this stage included pigeons, magpies,[6] dogs,[7] cats,[8] and some species of primate[9] like capuchin and rhesus monkeys.

In Piaget's Stage 6, the food is concealed without the animal seeing it – in other words, a screen is erected in front of one of the cups. You take the reward in your hand in such a way that the test subject can see it. But then you close your hand and put the reward beneath the cup behind the screen. Then you bring your closed hand into view once more and open it. Finally, you remove the screen and give your test subject the opportunity of finding the reward underneath the right cup. If you have hidden the reward beneath the cup while your hand was out of sight and so shown the subject that your hand is empty, an animal that can master this stage of object permanence will immediately start searching underneath the cup that was originally hidden from view. As far as we know, the only species that can manage this feat are the anthropoid apes, grey parrots, jackdaws, magpies and keas. We humans have this capacity from around two years old, which explains why we're always so confident of winning the 'simple' shell game.

If one can keep an object permanently in one's head in the form of a mental image, then one can also use that head to think. And if one can combine thoughts in a way that makes sense, then the phrase we use is 'logical thought'.

Logic

Anyone who has ever fed an animal will be familiar with this: you rattle a box or rustle paper or go to the shelf where the tins are kept. Almost simultaneously, your beloved housemate signals its interest. But is it necessarily logical thought at work when animals deduce on the basis of a particular sound that they are about to be fed? I'd like to say yes, but I'd be wrong to do so. What's going on here is conditioning. The pet has learned or rather has been conditioned to recognize that something tasty follows the corresponding stimulus (sound).

Rattling boxes are also deployed in experiments. Yet here the animals being tested are not permitted to be conditioned by the sound. Through their own deductive reasoning, they must arrive at the conclusion that a box that rattles has something in it, whereas an empty box makes no sound. That is logic in action, or to be more precise, an 'if–then' linking. *If* something is so, *then* it follows that something else must happen. But pay attention now, here's the most interesting part: an animal that can think logically and knows that one of two boxes is full can make the correct choice even if you shake only the empty box. According to the principle of exclusion, it must necessarily be the other box that contains food. By means of such experiments, anthropoid apes,[10] African grey parrots,[11] cockatoos[12] and keas[13] have so far been successfully tested for inferential reasoning. Humans can pass the test from the age of three years.[14] At first, dogs did not pass this test.[15] But in a subsequent experiment, three out of six dogs passed the test and even one out of six pigeons.[16] And in a study published very recently, all the dogs passed.[17] Of course, what happened here wasn't that the dogs miraculously evolved all of a sudden, but that the researchers became smarter. They devised their tests in a cleverer way by adapting them to dogs' behaviour.

Another test investigated the ability to recognize an unknown cause. Picture yourself sitting in the cinema. On the screen you see a dense forest. Off in the far distance, some trees are moving. There's not a breath of wind, and yet there's clearly something coming towards you. You can't see what's causing it, but the cracking sound of huge trees being felled is unmistakable – there must be a one-eyed giant approaching, or King Kong, or maybe a dinosaur from *Jurassic Park*. In any event, the movement spells danger for the film's hero. It's a wonderfully tense atmosphere. The question is whether other animals would be captivated by this scene. We enjoy experiencing a *frisson* of horror because we can imagine what kind of monster is capable of shaking treetops or causing whole trees to splinter and crack. We have the capacity to deduce a cause from an observed effect.

A test in behavioural biology is in much the same vein. Crows living in a large aviary with a food source are kept under observation. Next to the food source is a curtain, from which a stick protrudes towards the food. Two people enter the cage. One of them stands there in full view, stock still, while the second person disappears behind the curtain. Invisible to the crows, the second person now starts waving the stick around. For the crows, this is good reason not to fly down to the food source, as they risk being hit by the flailing stick and injured. But when the two people leave the aviary once more and the stick is lying immobile in its original position, the crows' behaviour instantly changes and they fly down to the food source.

There are only two possible explanations for this behaviour:

A: The crows make the association that an immobile stick, as opposed to a moving stick, is not dangerous (associative thinking).

B: The crows understand the connection between a hidden cause, namely the person behind the curtain, and an effect that spells danger for them (causal or causative thinking).

In behavioural biology, the simplest explanation is always the more plausible, and in the past this was how corresponding animal behaviour was always interpreted. But what happens if the experiment is altered ever so slightly? In this tweaked version, only one person enters the cage and stands in plain sight next to the curtain. But as if by magic the stick starts waving about menacingly. If explanation A is correct, then after the solitary, clearly visible person had left the aviary and the stick had stopped moving, the crows would immediately swoop down on the food again. Applying associative thinking, an immobile stick presents no danger. But in actual fact, the birds display a behaviour that can only be explained by option B.[18] They were nervous, often timid and only dared, if at all, to venture near the food source after intensive scrutiny of the stick and the curtain.

This experiment is particularly smart inasmuch as the situation is very close to natural living conditions. In the wild, branches do not move of their own accord either, and it is possible a danger may be lurking that first has to be ruled out. Even just a few years ago, only humans were credited with such behaviour; accordingly, observations of this kind were interpreted as purely associative thinking in the case of other species. Nowadays, the situation is markedly different, with an entire special issue of the renowned *Journal of Comparative Psychology* being devoted to this topic.[19]

Where dolphins are concerned, it has been known since 1984 that they are capable of making logical connections. Similar findings emerged from research into their language ability.[20] In order to gain these kinds of insights, one does not necessarily have to experiment with animals in captivity. Often all that is required is a trick to achieve similar results in the wild. For instance, ingenious researchers have collected fresh elephant urine and laid a slurry made of this urine and mud in the path of other elephants. They made sure that they took the urine of either familiar or unfamiliar elephants or that of animals whose position was either known to the test elephant or not. Try to imagine now what an elephant with a baffled expression looks like. For that is exactly the face an elephant pulls when it detects ahead of itself the scent of an elephant that it knows for certain is following behind it. In this way, the researchers tested first whether the animals knew where their associates and relatives were,[21] and second their logical grasp of the fact that an animal walking behind them could not possibly have peed in front of them.

*

Abstract thought

Thought becomes abstract when I am no longer dependent upon my original mental image. When I distinguish in my mind between different coins or images showing men and women, or pictures painted by Monet and Picasso, then I have created various categories. If I can now manage to distinguish not just between two categories but to assign totally new objects to a category, then I have begun to think abstractly. If you want to prepare yourself to apply for an assessment centre training programme, then you can find any number of such tests online. Why not try taking one of these tests, just for fun? If you do, you'll doubtless be even more astonished at the revelations in the following pages.

We have already seen how many capabilities only evolve over the course of an individual's development. So researchers were even more surprised to discover abstract thinking among ducklings. Lest you don't believe me, I should begin by giving you a little more background information.

You're all presumably familiar with the name Konrad Lorenz, the eccentric Austrian animal behaviour expert, who some eighty years ago succeeded in making a flock of geese so accustomed to his presence that they accepted him as their 'leader'. Even for those who hadn't heard of Lorenz, the release of the 1996 feature film *Fly Away Home* (which tells the story of a young girl teaching motherless geese she hatches from eggs to fly south for the winter) made many people aware of the bonds that such 'imprinting' forges. Lorenz's experiments were not just repeated by the film's heroine, Amy, but also by many other scientists. It has actually become a classic object lesson, in the same way as biology teachers in many schools conduct the 'mirror test' (see 'Self-awareness'; pp. 196–204) with fish and get their pupils to observe how aggressively they react to their own reflected image. It's really quite surprising when eminent scientists take up this experiment anew, as normally there's no credit to be gained from reprising old experiments.

Yet the Oxford behavioural ecologists Antone Martinho and Alex Kacelnik must have thought quite differently, since they hit upon the crazy idea of presenting ducklings with a test designed to ascertain one's ability to think in an abstract way. To pass this test, the test subjects, irrespective of whether they were animals or people, had to recognize that two spheres and two prisms comprised a category, whereas a cylinder belonged with a cone, and a cube with a rectangular solid. If a duckling is imprinted with two spheres directly after hatching, then that bird would also follow without hesitation two pyramids or two cones, or two cylinders. In the same way, a duckling that had been primed with two dissimilarly shaped geometrical solids would follow any other combination of two dissimilar solids. Now you might be thinking that this isn't a particularly impressive example of a creature showing the ability to think in abstract terms. When all's said and done, all that's happening here is that the duckling is following two geometrical shapes and thinks that they are its mother... that's not especially clever, right? Well, you might have a point there, but with the loss of flexibility one does gain the ability for absolute learning.

Once you've observed imprinting in action, you never forget it. But that's not really what this experiment was about – that was Konrad Lorenz's signal achievement all those decades ago. Here, it was all about discovering that the little duckling's brain was capable of recognizing a pattern in geometrical shapes. This pattern is abstracted, categorized and stored. Knowledge of this category put the little duckling in a position to identify completely new objects as belonging to that category and to act accordingly. We humans call precisely such a process 'abstract thinking'. And it is truly incredible that ducks are capable of performing such a complex operation, which explains why the famous journal *Science* saw fit to devote a couple of pages to Martinho and Kacelnik's research.[22] Ultimately, this study goes to show that abstract thought isn't really as complex as we once believed. Yet learning imprinting and the phase of life associated with it is essential for the survival of young animals. Eventually, the small ducklings have to be able to get by without the security of the nest, and there are many ducks on the pond: recognizing the mother duck therefore produces extreme evolutionary pressure and also holds great potential for evolution to find clever solutions.

In common with most songbirds, which can only learn new elements of song when they are young, ducklings can only think in such an abstract way during their first hours of life. The ability to do so is to hand when it is required, and is then switched off again. That works in roughly the same way that small children have no difficulty in beating almost every grown-up in memory games. As adults, we simply do not need to have so good a photographic memory.

Of course, the question now arises of whether other animals are also capable of such cognitive feats in their everyday lives. But watch out: in this case, it is not about possessing an ability in a particular phase of life. Many songbirds are capable of vocal learning when they are young, but all the same they could never speak a language, for as adults they would not be able to learn a single word. So the key point at issue is the capacity to think in an abstract way flexibly and throughout one's life. To state it clearly once more: we're talking here about abstract thinking – that is, the ability to reach a decision on the basis of an analogy, something that we really only deem humans to be capable of.

To find out whether animals have this ability, all we need do is play cards with them. The simplest way of finding out in the first place whether an animal can form categories is a game of Happy Families. My two sons are four years old and are gradually starting to find this game fun. But it took a great deal of patience to teach them even the basic rules.

A comparable effort is required if one wants to try to conduct such tests on animals. In the event of a laboratory animal recognizing a pair, it would be given a reward, and in this way the animal would understand what you wanted from it. But what if it's no longer just about recognizing an identical pair? What if it's about forming an analogy and deciding on the basis of that?

Imagine you are looking at a card with a blue square and a red square on it. There are two other cards next to it – one with a blue circle and a red circle; the other with a blue cross and a red square. Which of these two cards pairs with the first? The answer is both, since both have red and blue symbols on them; that would be a sensible analogy, but it applies to all three cards, and you are required to choose just one. OK, I hear you thinking, then I'll just pick the card that at least has one square on it. But that would be incorrect, as you ought to realize that it's all about choosing the card that likewise has *two identical shapes on it*. It's totally irrelevant what colour or shape is shown on the card, it's all to do with the analogy (analogical reasoning), and the fact that both things need to have the same basic configuration. So long as one is aware of that and knows the example, then it's all laughably simple, yet in my student days I had a part-time job helping out at assessment centres, and any number of applicants there failed because they hadn't prepared themselves for that sort of test and so didn't grasp what was required of them. Perhaps you should try it out for yourselves? Take a look at this scientific video in which a crow displays its powers of analogical reasoning.[23] It was only by the fourth example that I managed to be as quick as the crow. Until recently, only humans and the great apes were credited with this capacity.[24] Later came baboons[25] and then crows. It's exciting to speculate what other animals might be found to be capable of abstract thinking.

I don't want to deprive you of one last little juicy titbit of information: in this experiment, the crows drew analogies spontaneously. In other words, they hadn't been tutored beforehand to use analogies as a solution,[26] and would have moved on to the next round easily, unlike many of the candidates in my assessment centre. What's especially remarkable is the fact that the birds were also given a little treat even when they chose the wrong card. Even so, in most cases they still got the right answer.

Strategic thinking and creativity

At the beginning of this chapter, I took you to the town of Delphi in Greece. I was disappointed to find that the inscription 'Know yourself' no longer existed, and now I have to share another disappointment with you. Around 2,500 years ago, Aesop, a slave in the service of Ladmon of Samos, visited Delphi with the intention of sacrificing a considerable sum in gold coins to the gods in the name of his master. Aesop was no ordinary slave, but a scholar and moreover the writer of the famous *Fables*. As an intellectual, he may well have been looking forward to an erudite discussion with the learned citizens of Delphi, but in the event he was bitterly disappointed to find there nothing but stupidity and ignorance. It is said that he promptly refused to sacrifice the gold, whereupon he was accused of a serious crime by the enraged citizens and executed.[27] This action both robbed Greece of one of the cleverest and most creative thinkers of his age and disabused me once and for all of the illusion that Delphi was some special place. Yet Aesop's fame as the teller of fables still continues to this day, and in addition he may even have been the inventor of the first intelligence test.

Like every skilled experimenter, he wrapped up his test within a story. Are you familiar with the fable 'The Crow and the Pitcher'?

The story goes that Aesop once watched a thirsty crow struggling to get at the water at the bottom of a pitcher. After trying and failing, the clever animal collected some stones and dropped them into the jug, whereupon the water level rose and the bird was able to slake its thirst. Now, it is in the nature of these things that fables ostensibly tell the stories of animals but actually describe human virtues and foibles, and so this intelligence test has for 2,500 years remained a popular stratagem for distinguishing between clever and stupid people. Moreover, children in the Western world are capable of passing this test at the ages of around five to seven.[28]

Not long ago, the psychologist Sarah Jelbert from New Zealand had the idea of actually conducting this test with crows.[29] Aesop would presumably not have been at all surprised at the result, because in truth the laboratory animals mastered the test the researcher set them with no difficulty. They were given a variety of objects but ignored ones that floated or were hollow and selected instead heavy stones that would raise the water level. If they were given several water cylinders to choose from, they would always take the one with the highest water level. If the cylinder was filled with sand, they didn't even bother dropping stones into it – why would they? What might the crows have been thinking at that moment?

And yet their logic had its limitations. For instance, they couldn't distinguish between cylinders of different diameters. In a further test, the birds were presented with three transparent cylinders filled with water. The cylinder containing a food reward was too narrow for the birds to drop stones in, and they saw no reason to put stones into the other cylinders. What they did not know, though, was that the narrow cylinder was connected to one of the larger-diameter ones and that by dropping stones in that cylinder, they could have raised the water level in the narrow one and so got at the reward.

You might be thinking that was a really unfair test, yet human children can solve it from around the age of eight. Children of that age in the Western world have already experienced countless times how things can be manipulated indirectly. The light goes on when someone clicks a plastic object on the wall, and little cars can be manipulated perfectly by turning small wheels on a remote-control box. I'll bet you any money you like that we'll read a publication in the next few years describing how birds that observed comparable actions achieved very similar test results.

Yet I'll be totally frank with you and admit that these experiments with crows have also had their critics. Ultimately, it's all about determining whether the birds can devise a new solution to a problem creatively or whether they have arrived at the correct answers through trial and error. To find out, researchers proposed a somewhat modified experiment.[30] Unfortunately, I cannot tell you what the results were as the article was only published very recently and no doubt one or two years will elapse before someone takes up the suggestion and puts it into practice.

Similar tests, also involving raising the water level in a vessel, were conducted successfully with rooks[31] and great apes.[32] There is also a comparable test for dolphins, only this works in exactly the opposite way.

When conducting experiments with dolphins, it makes no sense to try to get them to drop stones into cylinders filled with water. The researchers therefore came up with the idea of installing a flotation panel inside a transparent cylinder; when it was loaded sufficiently with weights, this panel would sink to the base of the cylinder and release a fish as a reward. To begin with, a diver entered the tank and demonstrated to the dolphins the principle of how the device worked. He swam to a weight, picked it up and dropped in into the cylinder. The animals soon twigged that it took four weights to depress the panel and allow them to get at the fish. You would expect to observe this with all animals that can learn through imitation – already a considerable intellectual achievement that only very few species of animal are capable of. Yet the actual trick was as follows: the weights were placed roughly 40 metres (130 feet) away from the cylinder. This meant that the dolphins had to swim back and forth four times for 40 metres in order to get their reward. A strategically thinking mind would in this instance have picked up all four weights in one go and swum to the cylinder just the once. But that was by no means so simple, as many a master builder would no doubt attest: many workmen, especially those who are paid by the hour, prefer to go back and forth several times to fetch their materials and tools. They do this despite the fact that only one or two trips would suffice. As a building site manager, you'd be quite justified in asking yourself whether this was a case of cognitively underdeveloped individuals or whether, if your exasperation got the better of you, you weren't just being taken for a ride. Although the dolphins never saw another dolphin or a diver pick up more than one weight at a time, they quickly hit upon the idea of making their workload easier and began transporting several weights at once.[33]

Undoubtedly this was a creative solution and a clear example of strategic thinking. Yet all these comparative experiments only serve to highlight a problem. If one attempts to compare animals' performance with that of children of a certain age, it becomes problematic at some stage because one is not just testing cognitive abilities, but also at the same time the multi-layered experiences of a child in our complex world. In other words, the older children get, the more critical an eye we need to cast upon these comparisons, since one would never encounter an animal with remotely the same level of personal experiences as a six-year-old child. As a result, the animals are not capable of mastering complicated tests with anything near the same proficiency. I will have more to say on this topic in the chapter 'Experimental errors' (pp. 293–300).

*

For me, then, it is far more important to observe animals cleverly in the wild and to learn as much as possible about their repertoire of natural behaviours. Let us therefore turn our attention to a relatively easily observable behaviour, namely hunting or particular hunting strategies. Since we're discussing strategic thinking and creativity, normal hunting behaviour isn't the point at issue here. Rather, we're looking for behaviour that has come about through adaptation to particular environmental conditions. In actual fact, we're really only interested in short-term adaptations that cannot be explained through evolutionary mechanisms. Perhaps you've noticed in various wildlife documentaries how orcas, for instance, have a range of different hunting strategies. Thus they will circle round shoals of fish in order to drive them into a smaller space[34] or to force them to the surface;[35] they will beach themselves on flat foreshores to snatch basking sea lions;[36] or they will deliberately make waves in order to try to dislodge seals or penguins from ice floes.[37] Dolphins are similarly inventive, herding fish into shallower waters and then beaching themselves so as to be able to simply snap up the fish while lying on dry land;[38] producing a curtain of bubbles[39] or stirring up mud[40] to prevent fish from evading them; or using sponges as a kind of glove/snout protector while digging in the substrate for food.[41] To achieve a successful hunt, they will even cooperate with us humans: dolphins in both South America[42] and Asia[43] have been documented driving shoals into fishermen's nets and being rewarded for this service with a portion of the catch.

If these behaviours were just considered in isolation and only observed in a single species, we would readily be able to explain them as evolutionary adaptations. But in our examples, a species developed a number of different strategies, and there is much evidence to suggest that they pass these on to their progeny as a cultural asset. It almost seems as though a single dolphin once hit upon an idea and that all the rest have adopted it over the course of time.

It may be, though, that you are not the slightest bit surprised and expected to see such behaviour in dolphins. In that case, this next example will really give you pause for thought. Recently, French researchers were struck by some extremely unusual behaviour on the part of catfish. Catfish are the largest freshwater fish in Europe and can grow to well over 2 metres (6 feet) in length. The range of the European catfish (*Silurus glanis*) is mainly in Eastern Europe and only in the past few years has it been sighted more frequently in French rivers. To everyone's astonishment, in this new environment the catfish started to display a hunting behaviour that it has never exhibited in its home area, and which for that matter has never been observed among other fish in France. This involves the catfish propelling itself out of the water and – in the manner of the dolphins we encountered two paragraphs back – launching itself on to dry land to snatch seagulls. After just four seconds, it is back in its element, leaving human observers walking along the riverbank gaping in amazement. This technique has even earned European catfish the nickname 'freshwater killer whales'. As with the dolphins, its behaviour can hardly be explained by natural selection or evolutionary mechanisms.

In their paper on the subject, the French researchers don't go into the cognitive aspects of this behaviour, confining themselves instead to its ecological implications. This is the true focus of their research, since the European catfish is an invasive species, and was blamed by many anglers for the decline of native fish stocks until that claim was scientifically refuted.[44]

From a cognitive point of view, the behaviour of the catfish is almost unfeasible, and if it hadn't been so well and credibly documented, I for one would have taken it to be an angler's tall tale. Sure, there are some fish that will jump on to land momentarily to save themselves from predators, but that is an innate behaviour. And of course, the colonization of terrestrial realms by vertebrates has its origins in fish that began to climb out on to land to hunt. But that process took hundreds of thousands of years. There has also been some speculation as to whether the behaviour was already genetically fixed in the fish but was then blocked by the mechanisms of epigenetics.[45] But why should this blockage have been released exclusively in France?

The only remaining explanations are frankly incredible. How can a supposedly simple fish brain deviate from its innate reper-toire of behaviours, come up with something new practically overnight and improve itself through experience? And how has this behaviour managed to spread throughout the catfish popula-tion? Have all the catfish in France suddenly become little geniuses and thought up the trick for themselves, or have they learned from one another? I honestly don't know which explanation I find the more implausible. Try as I might, I cannot come up with any simpler explanation, and so I am forced to admit that we may have seriously underestimated the capabilities of fish. We will be returning to the subject of fish in the section 'Self-awareness' (pp. 196–204).

Now let's leave the water and look at some hunting strategies on land. Our relatives the chimpanzees are not vegetarians and are partial to meat. Like us, they are not perfectly designed predators, and so they have come up with various ingenious ways of getting hold of the animal protein they crave. In most people's eyes, the invention of the spear is confined to our hominid ancestors, and almost no one is aware that chimpanzees also fashion spears for hunting. Regarding the manufacture of these spears, the researchers talk of a methodical five-stage process.[46] Essentially, this involves the chimps stripping the leaves and twigs off switches and sharpening the tips with their teeth. For example, if a female chimpanzee discovers the den of a bushbaby, she will make one of these spears and instantly the supposed sanctuary becomes a deadly trap for the small and cute-looking nocturnal primate.

Alongside this skilled hunting technique, which can be successfully deployed by an individual animal without any great physical effort, there is also a cooperative hunting strategy involving an allocation of different roles and a complex execution. Unlike hunting in packs, in which each individual animal can take on different roles and functions according to the demands of the situation, chimpanzees are capable of planning a genuine drive hunt (battue). This entails four set roles:[47] one chimpanzee to flush out the prey; a group of beaters; a group of blockers; and the catchers. The usual victims of these coordinated hunts are red colobus monkeys.

Once the monkeys have been flushed out and driven by the beaters in a particular direction, the blockers suddenly appear. They bar the way, but in the narrow passage where an escape looks possible, the catchers wait quietly for their moment, and the trap is sprung. If you bear in mind that the hunt takes place not just at ground level but also in the third dimension and that the ambush has to be prepared quietly and unobtrusively, it becomes clear that this behaviour is in no way inferior to a human battue. The way in which the prey is handled after a successful hunt is also impressive – the spoils are divided fairly among all the chimpanzees that took part in the hunt, even though the scarers, beaters and blockers weren't actually in at the kill.

★

Mathematics

Let's stay with the Ancient Greeks for a bit. Perhaps you will recall the legend of Prometheus. He bamboozled Zeus with a trick and was severely punished for it. The trick was very simple: Prometheus knew that Zeus was a greedy skinflint and always took the largest portion for himself. Accordingly, he divided an animal he had sacrificed into two piles of meat. The smaller pile contained the choicest cuts, while the larger one comprised nothing but the bones, sinews, fat and skin. As anticipated, Zeus took the larger pile, thus showing that he knew set theory. He shared this capacity with most animals, as it is extremely practical to be able to tell what is larger. But when you consider that algebra, analysis, geometry and stochastic (that is, randomly determined) processes are also branches of set theory, you'll quickly appreciate what significance this basic ability has.

Over and above this, many animals can count, at least up to four. There is a nice article on this topic on Wikipedia,[48] so I'll refrain from listing all the many examples and experiments that go to prove it. But you might still enjoy the following anecdote.[49] At the Institute for Marine Mammal Studies in Mississippi, USA, a dolphin by the name of Kelly displayed some quite incredible behaviour. The dolphins there were trained to collect litter from the pool, and were given a small fish each as a reward. One day, rather than bringing it to the trainer, Kelly chose to hide a little scrap of paper she had found, possibly because she wasn't particularly hungry. That in itself is a remarkable achievement on the dolphin's part. Kelly had shown that she possessed a large measure of self-control. (The term used for this in scientific circles is inhibitory control, and in the normal course of events it is only ascribed to humans; we'll treat this subject further in the section on 'Theory of mind'; pp. 222–4.) In addition, Kelly displayed an understanding of time and made clear that she also grasped the theory of trade, since ultimately she was treating the scrap of paper like currency, which she could trade in for fish whenever she felt like it. But that wasn't the end of her creativity. No one had defined how large or small a piece of paper had to be to earn a reward, so Kelly came up with the idea of tearing the piece she had found into several smaller pieces so that she could claim repeated rewards. It seems incredible – but there's even better to come.

One day, she managed to catch a seagull and brought it to her trainer. He was thrilled and gave her several fish as a reward. So then Kelly hit upon a brilliant idea. Taking one of the fish, she positioned it in such a way that seagulls could easily get at it. Then she lay in wait, and was able to catch the next seagull and trade it in once again for some more fish. I'm just trying to think how many people, if you penned them in under similar circumstances, would behave as cleverly as Kelly?

The marshmallow test – on thinking about thinking

One morning I greeted my sons with the words: 'It's the gummy-bear test today!' They'd been looking forward to this test for days, as they love gummy bears more than anything. The original name of this experiment is the marshmallow test, as it was first conducted in the United States, where marshmallows are the last word in sweet treats for American kids.[50]

The test is very simple, and here's how I chose to run it: I gave each of my sons a plate with two gummy bears on it. They had to sit at a table and I held out the promise to them that they would each get a couple more bears when I came back in a few moments. There was just one catch – while I was out of the room, they had to hold off eating the two bears that lay there so invitingly on the plate in front of them, nor were they allowed to get up from their chairs. That's a real challenge for children who have only just turned four. Their thoughts and actions are still too geared to the moment and their self-control is not yet fully developed.

This inhibitory control, or impulse control or self-discipline that enables us to accept the concept of delayed gratification is an important way in which metacognition expresses itself. Only when I am able to judge my own knowledge, thoughts and actions can I reach decisions that do not simply arise directly from the moment. Patients with brain damage to their medial orbitofrontal cortex (mOFC) are no longer capable of delayed gratification.[51]

Yet the marshmallow test not only demonstrates the state of a person's cognitive development, it also allows far-reaching predictions about the future. It has been shown, for example, that children who pass this test at age four are more successful throughout the rest of their lives. It was even possible to show that this success was independent of factors such as intelligence or social environment.[52]

Much to my delight, my sons passed the test. On the other hand, I must confess I cheated. Though I left them to stew for the full prescribed fifteen minutes, there was one very important deviation from the original test. Normally, each child taking the test has to be left on its own to deal with temptation. In my case, there was an element of group dynamics in play, since it may have been the case that neither of my sons wanted to be the first to crack. Both aspects have to do with metacognition.

But what exactly is metacognition and how can it be tested? Of course, it's not possible to conduct any tests that involve language with animals. Metacognition is the process of gaining awareness and understanding of one's own thought processes and knowledge about what one knows, or conversely, thinking about what one knows or gaining knowledge of one's own thought processes. That's basically all it amounts to, and it relies upon the ability to be able to play conceptually with mental images and maintain a detached position of observation while doing so.

But let's strip this complex intellectual construct down to its bare minimum by doing another experiment. I take three transparent cylinders, into one of which – in full sight of my test subjects – I place a reward. Animals who are beyond Stage 4 on the Piaget scale naturally make straight for that cylinder and collect their reward. Now, in contrast to the experiment noted on pages 174–5, I obscure not one but all of the cylinders behind a screen. Then, unseen by my test subject, I place the reward in a random cylinder before removing the screen. Animals with metacognition know that they don't know which cylinder the reward was put in, and will therefore investigate all the cylinders, no matter how many of them there are. Animals lacking metacognition, on the other hand, will mostly choose the cylinder where they found the food reward previously. If one compares children aged two and a half with chimpanzees, there is no difference. Both are capable of passing this test.[53]

For the sake of argument, though, let's say this example is too simple for you and that you don't believe this experiment shows metacognition at work. In which case, I have another example for you. Quite out of the blue, you're required to give a speech. It's not really your subject area, and you know that mugging it up overnight won't be enough. Besides, you'd be too tired the next day anyway to be really on top of your presentation. Your boss, who is fortunately a nice guy, gives you the following options:

Option A: you accept the challenge and get the promotion you've been after for some years, along with a 40 per cent salary increase. The only catch is, if you mess up the presentation, you'll be fired.

Option B: you come clean with yourself and your boss and decline to give the presentation. You keep your position, but all

you'll get is a token salary increase of 2 per cent as a reward for your honesty.

This decision asks a great deal of you. In the balance are a pretty sizeable reward but also an associated risk, which you need to evaluate with the utmost care and precision in your present life situation. You might ask yourself: how quickly will I find a new job, can my family get by during this period come what may, and can I afford an interruption in my career? You mull over all possible aspects of your life and assess whether your aptitude and know-how is sufficient to enable you to deliver an outstanding speech.

Now imagine that you're a laboratory rat. Today, you're about to do a test that you've done lots of times already. The task is a straightforward one: you have to listen to two notes and decide whether they are of the same length. The test is a simple one if the notes are very clearly of different lengths, but it becomes much trickier if the difference is less than a second. You enter the test chamber and have to make your decision. The good thing about the experiment is that, after you have listened to the notes, you can decide whether you even want to take the test or not. And if you decide against, you'll still get a little titbit. On the other hand, if you take the test and pass it, you'll be rewarded with a whole heap of food, whereas if you try but fail, you'll get nothing. Rats only decided to take the test when they were certain of being able to tell the difference between notes; if the risk was too great, they declined to take part.[54] Without metacognition, the animals would not have been able to reflect on their own knowledge or ability and consequently would have been unable to take any risk-dependent decision. It's a wonderful, neat experiment and one that appeals to all animal behaviour researchers. Thus far, a wide range of animals, including the great apes,[55] dolphins,[56] rhesus monkeys,[57] capuchin monkeys,[58] pigeons[59] and even bees[60] have been successfully tested in experiments constructed according to this or a similar pattern.

Perhaps you're now thinking: but what about philosophical thought, or contemplation? Isn't that the supreme discipline of metacognition? Absolutely, but unfortunately contemplation by definition does not concern itself with external activities, but only with an internal dialogue. As a behavioural biologist, I therefore have no chance of testing this aptitude, for at the very moment in which action results from thought, it has ceased to be contemplation. But let me give you something to ponder as we leave this topic: who can say for sure that metacognition/contemplation isn't an ancient phenomenon devised by nature in order to dispel boredom? As soon as an episodic memory exists, it is possible to gain access to the contents of that memory and play around with them intellectually. Just ask yourself: when do you reflect on your own thoughts or in what moments do you indulge in contemplation? For me, it's often those moments when I have nothing else to do and am bored. It could be, therefore, that a guard dog indulges in more metacognition in a month than many a person does throughout their entire life.

23. Who am I, and who exactly are you?

Self-awareness

The simplest way of recognizing ourselves is by looking in a mirror. For this reason, ingenious researchers also came up with the idea of placing a mirror in front of test animals (MSR, or mirror self-recognition). The assumption behind this is that any creature that recognizes itself also knows that it is exercising an autonomous capacity for thinking and can in addition develop an idea of its own outward appearance.

The reactions to a mirror are extremely varied, but they can

broadly be summarized as falling into five categories. In the first category, the animals simply ignore the mirror. The image in the mirror has no meaning for them, though this does not necessarily mean that the animals do not possess any self-awareness. Perhaps it is just that the optical sense does not play such an important role for the animal species in question. Maybe it is the case that they would recognize themselves far more readily if they could smell their own bodily scent. It's possible that they would also recognize themselves if the sounds they customarily make were to be played back to them. You get the picture. In any event, the test is not necessarily equally effective for all animals.

In the second category, animals react to the mirror by displaying social behaviour. Frequently, this amounts to aggressive behaviour. For example, territorial fish react by attacking a mirror that has been placed in their territory. They will often do so to the point of complete exhaustion, and now we know about the 'endowment effect' (see 'Death cults and war' and 'Against reason'; pp. 158–65 and 231–41), we also know why this is the case. However, it is also possible for an animal to react affectionately to a mirror. In this instance, it will make unsuccessful attempts to get closer to its mirror image; such behaviour has often been observed in caged budgerigars.

Something remarkable occurs in the third category. Here, the animals clearly understand the concept of a mirror. For example, if one positions food in such a way that the animals can see it only in the mirror, then there are two possible ways they can react to the mirror. First, the animals will dash to the mirror, realize there is no food to be had there, look around puzzled, discover the real food source and rush to it. The second possibility is for the animals to see the reflection of the food in the mirror but to head directly for the actual food without bothering to make a detour to the mirror. In this category, the animals have a concept of what a mirror is and can call it up when required. This is a quite remarkable achievement, of which, for instance, pigs,[61] macaques,[62] African grey parrots,[63] crows[64] and several hundred breeds of dog[65] are capable.

Things take an amusing turn in the fourth category. Here too, the animals react at first by displaying social behaviour. Yet at some point they realize that something's not quite right. Frequently – and with evident astonishment – they grasp that they can control their counterpart in the mirror, and they begin to engage in what is called 'contingency checking', namely testing possible ways in which their opposite number might react. It makes no difference whether the animal concerned here is a small child or a great ape under the age of about two:[66] they wave their arms, legs and head about until it begins to hurt. After all, it's fun to have a remote effect; it's for precisely the same reason that we get a kick out of throwing stones or firing off rockets.

In the fifth and final category, so-called self-directed behaviour, the animals deal with their mirror counterpart in an unusual way. For instance, if you daub the forehead of a chimpanzee with paint, it will clutch at its head as soon as it sees the paint spot in the mirror. Normally, you would expect it to grasp at the paint spot on the forehead of its reflected image, since that is where it is seeing it. The fact that it actually wipes its own forehead shows that it is fully aware whom it is observing in the mirror and where the spot of paint is located.

In the first experiments of this nature, only chimps passed the test.[67] Of course, this provoked a storm of criticism at the time, challenging the finding that animals had self-awareness as implausible. Let's not forget that, at that time, in the 1970s, it was still hard for people to even get their heads around the idea that animals were capable of using tools. After all, it was still something of an article of faith then that tool use had made us humans into what we are today. Subsequently, the other great ape species, namely orangutans and gorillas,[68] also passed the test. Yet various other animals, including other species of primate, elephants, dolphins and parrots, failed it.[69]

Many of the criticisms were of course justified, and so the experiment was progressively refined over the years. Of necessity, a bird or a dolphin finds it very hard to touch its own forehead. In order to address this problem, marks were made on other parts of animals' bodies and it was deemed sufficient if the animals moved in such a way that they were clearly inspecting that body part. All of a sudden, other species began to pass the test, such as bottlenose dolphins[70] and orcas,[71] elephants,[72] Eurasian magpies,[73] and the playful keas[74] that we encountered on page 111. You can check YouTube[75] if you want to see exactly what such a test involves.

One further issue was that marking always entails touching, meaning that the behaviour displayed could just have been a reaction to having been touched. In order to counter this criticism, the animals were frequently touched but without leaving a mark. Thereafter, their reactions with and without having been marked were compared. If it transpired that the animals occupied themselves with the markings far longer and more intensively than when they had simply been touched, then researchers inferred from this that it really was the marking that triggered the behaviour rather than the mere fact of being touched.

Encouraged by the findings above, researchers retested animals that had originally failed. An interesting example in this regard is the rhesus monkeys. They do not belong to the great apes and do not have such a high degree of cognitive development. They did not pass the official mirror test, failing to react at all to a marking on their foreheads. But they did use a mirror to examine themselves very closely. One part of their body that they cannot normally see is their buttocks and sexual organs. If they are provided with a mirror, however, their fascination is unbounded, since this furnishes them with a new perspective.

The really interesting point is this. These animals can see the sexual organs of their fellow rhesus monkeys any time, so the fascination shown in the video[76] cannot be explained in any other way than that they are looking at their own genitals for the very first time. But for this to be the case, they must have understood that these are their own genitalia they are looking at. Therefore, if we bend the rules a bit, rhesus monkeys can also be deemed to have passed the mirror test,[77] despite not having taken any notice of the marks on their foreheads.

There are also cases, though, where the results of a supposedly successful completion of the mirror test needed to be reviewed. In an experiment conducted in the 1990s, pigeons also passed the test.[78] However, it turned out that they had been trained beforehand to peck at marks, which effectively invalidated the result.

Now things become really interesting: imagine that you did not possess any self-awareness. That's just not possible, you'll say, and I'd agree with you; I've also tried to do so and failed. But just consider this: how often do you think that you are conscious of yourself? Do you do so, for instance, when you barge into something and yell out 'ow!', or when you laugh at a joke, or when you're happy or are engrossed in your work? Of course not, even though you're not consciously noticing it yourself. It's only when you activate the capacity for self-awareness that this perspective becomes available to you. You can manage your daily life just fine without. What's more, who's to say how long we could stick it out sitting at a checkout in a supermarket if we had our capacity for self-awareness switched on the whole time? Who knows, perhaps we'd go crazy after just a couple of hours or days. The same goes for any monotonous task that we are required to carry out all too often in our day-to-day existence.

Two aspects concern me primarily here: we get along just fine on a daily basis without being conscious of ourselves the whole time, yet there is such a thing as 'a little bit' of self-awareness. Though it's impossible for us to imagine that, we can nevertheless observe it. For instance, among the elephants and Eurasian magpies mentioned above, who were tested for self-awareness by means of the mirror test, not all the animals passed the test. It therefore appears to be the case that this capability evolves in some individuals but not in others. It's a similar story with small children between the ages of eighteen and twenty-four months.[79] Some of them passed the mirror test, while others failed. Children in the Western world, where there are any number of mirrors, are more likely to pass the test than children who live in developing countries or in the bush and who are only rarely confronted by a mirror.[80] And if one tests children who have learned in advance to wipe a mark off the forehead of a doll, then these children are more likely to pass the mirror test than those who have not been prepared in this way.[81] This kind of test set-up is pretty much the same as that with the pigeons. We disputed the pigeons' capacity for self-awareness, yet we all agree that our children, who didn't play with a doll beforehand, nonetheless still have this capacity.

In the later chapter entitled 'Experimental errors' (see pp. 293–300) we will revisit such problems in greater detail. But the lesson here is that we must be very careful when claiming that other animals do not possess self-awareness. It could just be that we have not yet devised the right experiment. This must have been at the back of two French researchers' minds when they recently put forward the suggestion that, in the case of birds in particular, not only their optical sense but also their senses of hearing and smell should be taken into consideration.[82]

The following two examples – which I myself can scarcely believe – demonstrate that it even makes sense to investigate animals that no one ever suspected of having a capacity for self-awareness. It might appear perfectly plausible that highly developed mammals and even birds possess self-awareness, but what about fish? Although I studied marine biology and know quite a bit about fish, I would never have had the audacity to lay myself open to ridicule by conducting such an experiment. Some years ago, when I swam with manta rays off Western Australia, I had no inkling whatsoever that these majestic beasts gliding through the oceans might possess self-awareness. And yet manta rays have the largest brains of any fish, and in this respect it makes sense to initiate an experiment to that end. Although they failed to pass a mirror test under experimental conditions, they were found to examine themselves intensively in a mirror, just like the rhesus monkeys.[83]

If at this point your patience is at an end and your willingness to suspend disbelief baulks at the idea of self-aware fish, then I'd advise you not to read the following section but to skip to the next chapter.

If the results are correct and no researcher has been cheating, then the following observation far outstrips the most ambitious notions of what a ventral nerve cord system is capable of. Perhaps you can recall biology lessons at school. There, a rough distinction was drawn between the diffused nervous system of a jellyfish, the ventral nerve cord of insects, worms and spiders, and the central nervous system of vertebrates. In the simplest terms, one might perhaps say that one can use a diffused or ventral nerve cord to operate, while a central nervous system gives the animal the capacity to think. It is only the high density and spatial proximity of individual nerve cells within a central nervous system that facilitate more complex interconnections and hence make simple thought processes possible.

But how can I begin to convince you that even ants recognize themselves in mirrors and wipe marks off themselves in just the same way as chimpanzees?[84] In contrast to most other animal species tested, among which it was frequently the case that not all individuals successfully completed the mirror test, almost all the ants wiped off the marks that had been put on them and so passed the test with flying colours. What's more, it turned out that, just like human children under the age of eighteen months, young ants had not yet developed this capability. I fancy I can hear something of the researchers' own incredulity when I read their notes on this experiment; also, they are at pains to point out that self recognition does not equate to a state of being aware of oneself.

To try to make the development of such a faculty at all plausible, the researchers adduce the fact that ants customarily react extremely aggressively to ants from another ant colony. An ant that no longer looks the same as the rest therefore runs the risk of being ostracized from its social group. This explanation doesn't completely convince me, however, for what ant has a mirror to hand to quickly check its appearance before returning to the anthill?

Yet these observations begin to make sense to some extent if you have anything to do with robotics. This subject area was a complete mystery to me until I was invited to contribute a chapter on the interaction between humans and dolphins to a book.[85] This volume also included a chapter entitled 'The significance of body and form in interactions with social robots'. It is extremely useful for a robot to have the capacity to recognize itself. For instance, an awareness of its own form and size makes it easier for it to navigate or to manipulate its own 'limbs'. In robotics technology, great store is set by the quality of self-management, which is only made possible through self-awareness. Of course, one can programme a computer to recognize its exterior appearance. But that's not the desired goal; instead, the aim is to get the computer to do this for itself.

In another test, which works slightly differently from the mirror test in that it relies upon a verbal instruction, the only test subjects able to provide correct answers are those who are aware of themselves. Hard though it is to believe, a commercial 'Nao' toy robot manufactured by the French robotics company Aldebaran Robotics actually passed this test. The remarkable thing about these little humanoid robots is their ability to speak and to understand human speech. In addition, they can also be simply programmed. For instance, three of these funny little guys were told that two of them had been given a 'dumbing pill' that prevented them from speaking. Then all of them were asked which of them could speak. And lo and behold, the voice response of two of them really had been incapacitated. The other one answered correctly 'I don't know!' Immediately afterwards, it revised its statement and said: 'Sorry, I know now. I was able to prove that I was not given a dumbing pill.'[86] After all, it was the only one that could speak.

Considering the limited processing power available within the computer or the ventral nerve cord, both results are impressive, and yet neither ants nor robots possess self-awareness. Nonetheless, they are indisputably capable of recognizing themselves. A true capacity for self-awareness is part of the so-called theory of mind – in other words, the notion that there exists one's own Self along with other Selves. But more on this subject in the chapter 'The thinking apparatus' (pp. 241–50).

Personality

Our personality is nothing more and nothing less than that which we are, that which we perceive, and how we appear to others. Our personality changes over the course of our lives, but by and large we remain true to it. We even have a word that we use to encapsulate the complex entity that is our personality – our name. Even if I have a tax or passport number that identifies me clearly for administrative purposes, my name is what identifies me as an individual. Now, this naturally begs the question whether names or personalities also exist within the animal kingdom. Most people who share their lives with pets would answer this question with a resounding 'yes'. For them, Fifi has a totally different personality from Lucy, and there's no comparison between tomcat Buster and sweet kitty-cat Rosie. They're given names like us humans, which they respond to throughout their lives. However, it's not as simple as that, and so we need first to take a brief journey back in time.

Some sixty years ago, husband and wife researchers David and Melba Caldwell thought that they had made an incredible discovery. They were studying dolphins that had recently been caught for dolphinariums and determined that the animals were clearly calling to other dolphins with a whistle. At the time, a call like this was regarded as sensational, since animals were not thought to be capable of such things. It transpired that whenever these animals were separated from their group members, they began to whistle like mad. It was therefore logical to suppose that these whistles were intended to summon others, possibly to help. Scientists thus began to refer to a 'contact or meeting call', in other words a call promoting cohesion within the group. But when the researchers studied the whistles more closely, they found that each animal produced a different whistle. That was something of a mystery, for what use was it if every animal called in a different way – after all, other dolphins couldn't be expected to pay heed to every individual contact call uttered. The whole point of a contact call was precisely that it should be a particular type of call that meant the same thing to every animal that heard it. After pondering the matter, the Caldwells came up with the idea that the calls might possibly be identification whistles.[87] But that was a bold hypothesis and even more implausible than that of the contact call, for what would be the purpose of that?

As we have already learned, some animals are capable, just like us humans, of identifying the caller from their voice. Therefore, there would be absolutely no reason to have a special identification call if one could be identified anyhow from one's normal calls. And yet we humans still have identification calls, in the form of our names. These are extremely useful when it comes to summoning someone, enquiring after someone, or talking about a person who is not present: 'Hey, Chris, come over here!' 'Any idea where Gerald is?' 'I was round at Richard's yesterday having dinner.' This would be a real challenge if Richard didn't have a name and we had to imitate his voice instead.

Names are therefore a very practical thing; we couldn't imagine people's everyday lives without them. But do animals likewise really have names? After many years of debate, this now seems the most plausible explanation. Let's take a look at everyday life among dolphins, for example.

Among dolphins, it is extremely impolite, indeed almost an aggressive gesture, to swim directly towards one another. In complete ignorance of this rule, and working from the assumption that we humans can do what we like, many years ago I swam straight at a group of dolphins that were swimming on the surface. From the point of view of the dolphins, that amounted to throwing down the gauntlet. There were two possibilities for how to respond: to turn away and back down, or to maintain their course. I imagine you'll be able to guess what happened. The dolphins did not regard me as dominant, and I didn't dare to swim to one side and present my flank to them. So I pulled in my stomach and made myself quite flat. But at that very moment, the dorsal fin of the middle dolphin slammed into my sternum, knocking all the wind out of me. Maybe you're thinking that was an accident and that the dolphin inadvertently failed to dive deeper underneath me. That wasn't the case, though. Dolphins have little pressure sensors that they use to pick up vortices of water and dispel them by means of tiny muscle movements. This makes it easier for them to swim faster, for in the absence of vortices, their hydrodynamic drag is reduced. Yet this same sensitivity also protects them from incurring injuries and collisions. As such, it is practically impossible to touch a dolphin underwater if it doesn't want to be touched. Conversely, any contact that is made is quite deliberate; I was being made to pay the price for my incivility.

In our human culture, people greet one another quite differently. We seek eye contact, go straight up to the person that we want to say hello to, extend our hand in greeting and introduce ourselves. By contrast, a polite dolphin will approach another in friendly fashion obliquely and from behind, calling out its identification whistle a few times and then pausing so that the other dolphin can respond. If the other dolphin answers by giving its own identification call, the invitation has been settled and they swim alongside one another for a spell.[88] So, humans and dolphins greet one another in thoroughly comparable ways: namely with the aid of elements of body language and by mentioning their own name.

As a second example, I'd ask you to picture the following situation: you are walking across a broad expanse of snow when fog suddenly descends. You can only see a few metres ahead, and lose sight of the friends you were with. So you call your friend's name. This situation is broadly akin to the conditions underwater. You can't see especially well, but calling works just fine, since water conducts sound better than air. For instance, if you were in water at a depth of a couple of hundred metres and could let off a firecracker, you should – according to where you're located – be able to hear the bang again around three hours later, by which time the noise of the report will have travelled round the entire globe.[89]

But to return to our names: the friend you have called out to in the fog would customarily reply by saying 'I'm here!' It's much the same with dolphins. If you're looking for a particular animal, you call him or her by name. You don't get the response 'I'm here!' but instead the name you have called. In such a situation, then, my son would call back to me: 'It's Vitus here!' – and in the same way, a dolphin replies by giving its identification whistle.[90] As far as I know, this is unique in the animal kingdom. There have even been some observations of an identification whistle being used without the corresponding animal being in the vicinity.[91] If one considers that animals have shown under experimental conditions that they have no problem learning and employing particular signals for particular objects,[92] and that over and above this they are capable of mastering grammatical rules,[93] then one might reasonably suppose that the animals are talking about the absent dolphin – though this is pure speculation thus far.[94]

Another endearing way in which dolphins use their identification whistle involves the 'marriage' of male dolphins. Yes, you read that right – male dolphins are gay and get married. For the most part, after puberty they live together in stable groups of two or three. And in the same way that human couples give themselves a common surname, so the identification whistles of married male dolphins converge and come to resemble one another.[95] There could hardly be a clearer way of indicating that they are living together.

It may be that we will in future be able to observe a similar phenomenon among parrots. It has recently emerged that these creatures also have individual calls and, like dolphins, use them to call and announce their presence to one another.[96]

A journalist once asked me whether that was also the case among pigs. She had done a piece on pigs in state-of-the-art piggeries being summoned to the feeding trough with an individual signal. Only the pig that was called would have its food dispensed into the trough. This simple trick had served to lessen the incidence of argy-bargy and aggression among the animals at feeding time. So, you might think they are taking the signal to be their name, but in actual fact they were simply being conditioned to respond to the signal. They had learned that the individual call meant food exclusively for them. Ultimately that was no different from when a dog or a guinea pig responds to the name we have given it.

It's a quite different situation among dolphins. As juveniles, they start off by imitating the identification whistles of their mothers, and over the course of the first few months of their life change it into their own identification whistle.[97] In this way, they create their own name. For some years now we have even known that the acquisition of individual names follows a very similar pattern among green-rumped parrotlets (*Forpus passerinus*), though in their case the juveniles take their cue from both parents.[98] This is impressive, especially when one considers that human children are not capable of this, but instead get their names from their parents and use them exactly in the form they learned them. Yet we humans have an advantage in that we can change our names any time we like, every few minutes should we choose to do so. From one moment to the next, I can style myself Hermann Hesse and the following day call myself Frank Schätzing. Admittedly, nobody would actually do it, because it would make things too confusing, but we have the capacity to do so, and it is highly doubtful that animals would be capable of doing the same. One exception to this is the 'married' dolphins.

Now things get a bit complicated, because our use of language lags behind the findings of the most recent research. There are many current researchers who actually proceed from the assumption that certain animal species have evolved to the same level as humans. The buzzword here is 'personhood'. Generally speaking, this denotes an animal that does not differ fundamentally from human beings in its basic cognitive faculties. These, then, are animals with self-awareness and empathy, a lifelong memory, or more precisely a biography and an understanding of space and time, together with the capacity for strategic, systematic and logical thought, a simple language with grammatical rules, a culture and a conception of fairness and a sense of justice. This is clearly a huge topic with far-reaching ethical and legal ramifications.[99]

At this point, the concept of personality should come into play, and indeed this is now something utterly different, although until very recently people were of the view that only humans possessed a personality and most people made no distinction between a person and a personality.

The term used in biology when talking about personality is 'consistent individual differences' (CIDs). This refers to, for example, 'temperament', 'nature', 'character' or 'individuality'. To gain a better understanding of this, we should make a brief excursion into the human psyche. A relatively recent branch of psychology, personality studies, concerns itself with characteristics in human behaviour that vary from person to person, despite the fact that these individuals all experience the same situation. Thus, for instance, certain people react phlegmatically and others aggressively, and in the main, these patterns of behaviour remain the same for a long time or even for a person's entire lifespan. It is as if our personality has been assembled according to a very specific recipe.

A possibly somewhat lame analogy makes clear how researchers have approached this question. Over the course of our lives, we have all eaten countless meals. In principle, each individual meal differs at least slightly from the others. Sometimes the fried potatoes are firm and waxy, sometimes more floury, sometimes they have been fried in olive oil from Italy or from Greece, and sometimes they have had salt or dried thyme, or both, added. Natural foodstuffs display a fantastic variety, and quite rightly we regard great cooks as true artists, who can play around with ingredients and create new taste experiences. And yet all taste sensations can ultimately be reduced to the five primary tastes that our senses are capable of registering: sweet, salty, sour, bitter and umami.[100] We cannot taste anything other than these, and the great variety results from combining them. It's a similar situation where our personality is concerned. To this end, researchers have collected and grouped together hundreds of human personal characteristics. What has emerged are the so called Big Five key personality traits, to which all other personal character variations can be assigned. The following table has been firmly established in psychology for around twenty years, and has been repeatedly borne out by thousands of individual studies:[101]

Factor	Factor definition	Weakly present	Strongly pronounced
Openness	Toleration for and exploration of the unfamiliar	Conservative, cautious, reticent	Inventive, inquisitive, imaginative, adventurous
Conscientiousness	Degree of self-control, exactitude and motivation in goal-directed behaviour	Carefree, negligent	Effective, organized
Extroversion	Quantity and intensity of interpersonal interaction	Reticent, reserved	Gregarious, capacity for enthusiasm

Agreeableness	Level of altruism and cooperation	Competitive, antagonistic, aggressive	Cooperative, friendly, empathetic
Neuroticism	Individual variations in the experiencing of negative emotions	Self-assured, calm, content	Emotional, fragile, apprehensive

Let's look, for example, at inquisitiveness as one aspect of openness. In our modern culture, inquisitiveness is seen as something desirable. This was not always so; one need only think of the old dictum 'curiosity killed the cat'. The past 2,000 years, conditioned by religious obedience and piety, represent instead a conservative outlook. And indeed, from a biological perspective, this is actually very sensible. If you behave in a conservative and reserved manner, then you exercise caution in the face of change and take care to preserve conditions that are clearly successful. Striking out in new directions entails enormous risks. If you enter a new terrain, for instance, you may not find enough to eat. If you change an established hunting strategy, then you incur the danger of coming away empty-handed and even getting injured. Yet why should curiosity even exist? Curiosity is the root of change and hence is a driver of evolution. If people had not been curious about what happens when you strike a stone against something, for example, there would have been no Stone Age and presumably humankind would not have evolved.

Curiosity can also be extremely useful in everyday life. If I belong to the select band of people within a population who exhibit curiosity, then it is highly probable that I won't always do what everyone else does. On the whole, that is not to my advantage, since I will tumble from that promising-looking rock ledge, or my boldness will simply lead me to become a tasty meal for a predatory animal. But occasionally I might strike lucky and the new source of food I find will benefit the entire population. This furnishes us more or less with an explanation of how the great spectrum of different human personality traits came about. There is no such thing as a single optimal strategy or perfect adaptation. The risk that external changes could bring about a catastrophe would be simply too great. Stability – that is, long-term survival – resides rather in diversity and dynamism.

So it will come as no surprise now when I tell you that evolution did not come up with this trick only with the emergence of the human species.[102] Indeed, over the past decade or so, diverse personality traits have been shown to be present in countless species of animal. Personality is not confined to cognitively highly developed animals. There are hermit crabs that emerge from their shells far more frequently than others, or widow spiders that choose their occupation on the basis of individual temperament. What's that, you say – spiders have occupations?! Yes, there really is a species of social spider in which the more aggressive individuals perform the tasks of food acquisition and defence, while the more mild-mannered individuals are responsible for bringing up the young.[103]

On account of the huge significance of individuality not only for an understanding of behaviour in humans and animals alike, but also for the evolutionary process, in 2015 a working group of the Society for Experimental Biology met in Prague and discussed, among other things, different personality traits among hermit crabs, rainbow trout, zebra fish, starfish, bullhead sharks and kangaroo rats.[104]

It is hardly surprising, then, that one dog should strike us as being unlike another – quite simply, it has its own personality.[105] Dog owners who in the past were accused of anthropomorphism and naivety in describing their pets as friendly or inquisitive have now been vindicated. All they were doing was noticing what science has only latterly been able to confirm through a combination of wide-ranging experiments and genetic studies.[106]

This brings us directly on to the next question. To what extent are our personalities shaped by our experiences, and how much of our personality was genetically preprogrammed into us at birth? This latter factor varies from species to species. According to the findings of research into twins, among us humans nature rather than nurture accounts for between half and two-thirds of our personality.[107] Put another way, only a third to a half of your personality is under your control.

But let's return for a moment to the aspect of openness and curiosity and willingness to explore. It has been possible to show, for instance, that among great tits the urge to discover new things correlates with dopamine receptors in the brain.[108] At this point, I refer you, as so often, to the upcoming chapter on 'Sentimentality' (see pp. 269–82), where the effects of dopamine will be discussed further. Now, the architecture of these receptors, just like every other component of our bodies, is encoded within our genes. Nature employs a little trick here whereby it can control the number of receiver antennae (receptors) for dopamine. For this purpose, small methyl molecules (CH_3) are attached by means of enzymes to the DNA packaging proteins. These in turn ensure that our DNA is so firmly knotted that the information can no longer be read from it.

We humans have around 2 metres (6 feet) of DNA strands within a single set of chromosomes. The DNA does not simply float about in our cell nuclei but is wrapped around small balls of protein known as histones. If a methyl molecule is attached to these, they attract one another and so impede the readability of the DNA. This process of so-called methylation is of such decisive importance for our lives that a separate branch of science is now devoted to the effects of these molecules and other mechanisms. This is known as epigenetics, a branch of genetics still in its infancy that seeks to explain why, for instance, the tree that grows from a seed of the Scots pine (*Pinus sylvestris*) in Germany and other parts of northern Europe looks very different from the tree that grows from similar seed in the Mediterranean region.[109] Epigenetics explains the effect of environmental factors on genes, or on which genes are selected under what conditions. Thus, a Scots pine in Lombardy looks quite different from a Scots pine growing by Loch Lomond. Nonetheless, it remains the same species with identical genes; the only variable factor is where you plant the seed.

It is not only external appearance that is shaped in this way. Internal processes and behaviour, such as our character, for example, are also subject to these mechanisms. Thus, if experience tells me that curiosity is worthwhile in my life, then within the nuclei of my brain cells, the 'histone knots' around the genes responsible for the architecture of the dopamine receptors are untied, and in consequence I become even more inquisitive. On the other hand, if I have bad experiences with this character trait, the DNA will be knotted afresh. Research on this aspect of genetics is still very much in its infancy, and of course I have grossly simplified the whole complex context, but at root this same mechanism can be presumed to operate in the same way among us humans as it does among great tits and other animals.

Fortunately, we are not helplessly at the mercy of these mechanisms. We perceive their workings as feelings, and as we have said repeatedly, the more highly developed our brain is, the greater its capacity to override the power of feelings – that is, the effects of hormones and neurological messenger substances. At this point, our memories and experiences or conditionings from childhood come to play an important and possibly even dominant role. Yet there is no denying the influence exerted by epigenetics through methylation and other mechanisms of gene regulation, either; this may even be largely responsible for shaping our personality. Is that really so incredible?

If one compares the methylation of chromosomes in two sets of identical twins at the ages of three and fifty, it soon becomes apparent how strongly we are subject to the influence of epigenetics.[110] Whereas the chromosomes of the three-year-old twins are as alike as two peas in a pod, the chromosomes of the fifty-year-olds look totally different. Environmental factors and our experiences have changed our chromosomes and hence also us and our behaviour.

Here's a small tip for you: people who eat lots of bananas are providing their bodies with ample methyl molecules and thereby enabling themselves and their offspring to adapt better to their environment and to lead a more healthy life.

I'll admit that the last two chapters have been pretty strong stuff. They chip away at our egocentric worldview and our image of ourselves. It takes some time to digest all that. The individual stands front and centre in our Western culture. Be it religion, philosophy or psychology, almost everything turns on the development of the individual person or personality.

Yet if the simplest organisms were found to possess a personality and our cherished Self could be explained in large part through genetically preprogrammed modes of behaviour, where would that leave us humans? How are we to think of ourselves on learning that personal predilections may well be just as important for spiders in their choice of occupation as they are for us? What do we make of the fact that we share these personality-forming mechanisms with great tits, for example? What do a fish or an ant feel when they see themselves in a mirror? How highly developed are we if we cannot even begin to imagine what self-recognition or self-awareness feels like?

It's a sobering thought and seems illogical, but if we really want to understand ourselves, we need to turn away from ourselves. The real mental achievement resides not in our preoccupation with ourselves but in our interaction with others. Here, then, we arrive at the 'theory of mind' and our ability to empathize with others. Let us therefore examine this quite astounding capacity to simulate the thoughts and feelings of others in our own thoughts and on this basis to make predictions about their future behaviour. It's this capability alone that enables us to cheat others or to offer them sympathetic support.

I know that you exist

For many people, sympathy is one of the most important and noble qualities that hallmarks us humans. It feels good to help someone else and to find that this help is gratefully received. But are there also some animals who not only register the distress of others but help them as well? My study is on the ground floor of our house; I have two large French windows that give out on to the garden, and these are mostly left open during the summer. Of course, now and then a field mouse will stray into my room and make itself comfortable there. However much of an animal lover I am, I don't want these cute little rodents in the house, so I got hold of a live trap. I don't need to use it a lot, but when I do it works a treat, since I fill it with a hazelnut spread whose name I forbear to mention here. It works every time without fail.

One time last summer, I found that the trap had been sprung, but there was no mouse to be seen, nor any hazelnut spread for that matter. What on earth had happened? I had hidden the trap under my staircase in a partition wall where some building materials were stored – a paradise for mice. When I checked the trap in the evening, I found that all kinds of little stones had been wedged from outside under the flap of the trap. At first I couldn't get my head around what I was seeing there. Hadn't I set the trap, and had the stones perhaps slipped into the trap while I was setting it up? That made no sense, but what was I looking at here? Had a mouse got into the trap and been freed by another mouse? Had the second mouse recognized the gravity of the situation and realized that you could easily push the trap open from outside, but that this kept falling shut again? So had this second mouse then worked out that it could push stones under the flap to prevent it from falling shut? And finally, had this second mouse succeeded in freeing its companion? I was again reminded of Sherlock Holmes, but I couldn't really believe my own story. It seemed quite inconceivable to me that mice might be capable of such a complex procedure, so I started to do some research.

A quick browse under the search terms 'prosocial behaviour' and 'empathy' soon yielded some results from the pages of the renowned journal *Science*. It did indeed contain a report of an experiment in which rats had freed other rats from a trap. They even did that when the trapped rats weren't released into their communal enclosure. This last point is particularly significant since it underlines the fact that this was a purely altruistic – that is, a totally unselfish – piece of behaviour. The rats who freed their conspecifics genuinely had nothing to gain from their efforts. But even better was to come: the researchers presented their rats with a second cage that they could choose to open. This one did not contain any fellow rats to be freed but instead the most alluring temptations in the form of chocolate-covered crispy cereal. What I then read made me see my empty mousetrap with fresh eyes. The rats chose to open the trap containing the other rats first before they then all joined together to make free with the second cage containing the chocolate.[111] Here, then, I had found my proof that rodents, or at least rats, were capable of recognizing and manipulating complex mechanisms. Not just that, however – they did so out of empathy, without receiving any reward. I would be prepared to bet my bottom dollar that mice would also pass this test – or certainly the smart little creatures who lived in my garden, in any event.

'Clever Hans'

In the early years of the twentieth century, it seemed that a dream had been realized. Science and technology were seen as the great saviours of the future, and people began to look at nature and the animal world around them in a totally new light. It chimed admirably with this prevailing mood within society that horses should be thought capable of reading and counting. And so it was that a German mathematics teacher by the name of Wilhelm von Osten and his horse, Hans, attained cult status during this period. 'Clever Hans' (*Der kluge Hans*), as he came to be known, could count and read. One could even ask him how many men in the room were wearing straw boaters. So, it appeared that he was numerate and could in addition distinguish between men and women and between those with or without straw hats. He could also remember people's names and recognize them on photographs even after several years.[112] It didn't take long for suspicions to be voiced that little tricks were being employed to give the animal a series of surreptitious commands. But even total strangers were able to get Hans to count and read, thus precluding the possibility that his trainer was coaching him.

Critics of this astonishing achievement by an animal even postulated that Clever Hans might have the gift of telepathy. In other words, they were more prepared to believe in telepathy than in animal intelligence. But what no one at the time suspected was that the sceptics were actually very close to the truth; we will go into greater detail on this point presently. In 1904, a scientific commission was set up under the chairmanship of the then-celebrated Berlin researcher Carl Stumpf, with a remit to either confirm the sensation or to expose Clever Hans and his owner as con artists. The sources are somewhat contradictory, but apparently it came to pass that, after the commission had been all but convinced of the animal's intelligence, a student called Oskar Pfungst, having conducted some scientific experiments of his own devising, hit upon the idea of covering the animal's head with a blanket. And lo and behold, all of a sudden Clever Hans had nothing to say, because he could no longer see the person facing him.[113]

Notwithstanding this insight, Clever Hans' career continued to thrive, and he was even taken on by the businessman Karl Krall after the death of his first owner, the mathematics teacher. Krall founded a regular 'stable for the purposes of instruction' and published his findings in the book 'Thinking Animals: Contributions to animal psychology on the basis of my own experimental observations'.[114]

Even today, the experiments carried out by Oskar Pfungst are celebrated as a scientific breakthrough, compelling scientific luminaries of the time, such as the German 'Indiana Jones' Carl Georg Schillings,[115] to openly admit that they had been wrong and to stress the importance of applying rigorous scientific methodology to experimentation. However, the findings of the commission represented a disaster for behavioural biology. It was many decades before another scientist was able to engage with the question of the intellectual capabilities of animals without running the risk of becoming a laughing stock. Evidence of the intelligence of animals was regarded as a confidence trick, and the primate researcher Wolfgang Köhler had to wait more than thirty years for his studies into tool use by chimpanzees to be given the recognition they deserved.

The reason for this was the inability properly to evaluate the achievement of Clever Hans or other animals. The wheel had been invented, so to speak, but no one had thought to build a cart.

Yet what was it exactly that was thought to be so special about the horse's achievements? Clever Hans was capable of reading unconscious signals conveyed through tiny hints in the facial expressions and the body language of the person asking him questions. It would appear that he was even better at this than we humans are, otherwise we wouldn't have to go to the trouble of sending our managers, diplomats and psychologists to seminars to tutor them in the art of non-verbal communication. By closely observing his questioner, even when the latter made every effort to stay neutral, Clever Hans was able to tell with the accuracy of a lie detector when he was giving the correct answer. This ability can be explained in terms of two remarkable skills:

- Mind reading
- Behaviour reading.

Perhaps you can still recall our earlier section on 'Body language and pointing' (see pp. 95–9). I was surprised to learn that horses behaved in exactly the same way as my dog. Furthermore, they also tried to make contact with their keepers and as soon as they managed to make eye contact, they would quickly look down at the bucket with the fodder that was out of their reach.[116] Horses therefore showed an aptitude that makes it clear that they, like us, inhabit a communally shared world. But had they actually displayed mind-reading or behaviour-reading capabilities? And had Clever Hans engaged in a measure of category formation (see the sections on 'Mental images', 'Logic' and 'Abstract thought';

pp. 171–83) that exceeded our human capabilities? Had he analysed and categorized the most diverse expressions of action in people that were strangers to him and recognized patterns of behaviour that all people possess in equal measure (behaviour reading = reading body language)? Had he simply registered these and then used his hoof to tap out the correct number? Or had Clever Hans been capable of empathizing with another creature and in so doing read its feelings and thoughts (mind reading)?

Theory of mind

We humans are certainly capable of doing this. The underlying cognitive skill, which recent studies based on the most rigorous research methods still assure us is exclusive to humans, is called theory of mind and is, in my opinion, a trailblazer for the evolution of the intellect and for complex social behaviour and culture. It was in 1978 – in other words, more than sixty years after Clever Hans – that it was posited for the first time in animals.[117]

But first things first. What exactly is the theory of mind, how can one test it, and which animals have the capacity for it? Thus far, no clear definition exists for the theory of mind. But perhaps it may be summarized broadly in the following terms: if I have a conception of mental processes or states such as knowledge, desires, needs or beliefs, then I have a theory of mind. In philosophy, the term is often used jointly or even analogously with the term metacognition. According to this, our rats from the section 'The marshmallow test' (pp. 192–6) would possess a theory of mind, since they were aware of their own state of knowledge.

Equally, one could argue that a horse that waits until a person makes eye contact with it and only then looks down at its feed bucket must have an awareness of the inner mental state of that person. The horse may possibly be aware of the fact that it is only attentive once eye contact has been established. Attentiveness is an internal state. But then a smart colleague of mine alerted me to another possible explanation. The horse might simply be reacting to the person's behaviour, namely to the direction they were looking in. One would therefore need to test whether the horse would also react in this way to someone who was blindfolded. If they recognized that a person wearing a blindfold could not respond to them, and consequently did not bother looking down at the feed bucket, then they would clearly have a conception of the inner state of people.

The zoologist Marthe Kiley-Worthington, who was born in England but grew up in Kenya, is currently head of the Eco Etho Research and Education Centre[118] in France, which is concerned with matters of animal welfare. She believes that all mammals, from the smallest mouse right up to the elephant, have evolved a theory of mind. In her opinion they are therefore capable of empathizing with other creatures and of drawing their own conclusions from this insight.[119] Mice, for example, react with panicked flight when they hear another mouse emitting stress cries as a result of being subjected to electric shocks. Because genetically different strains of mice display reactions of differing intensity, the scientists were able to rule out the possibility that this was simply a learned response.[120] But how exactly can one explain the varying behaviour of the genetically different mouse strains? If one considers the results from metacognition experiments, then it would be perfectly possible to posit that the animals are more or less capable of empathizing with the stress and panicked state of other mice. I am very excited about what future findings scientific research in this field might yield. Yet whenever I think of this experiment, I cannot help but call to mind Douglas Adams' novel *The Hitchhiker's Guide to the Galaxy*. In it, mice allow themselves to have experiments conducted on them in order to test how intelligent and sympathetic the experimenters are. There's no doubt what their judgement would be of the researchers who created stress in mice by giving them electric shocks.

False belief

However, let's leave mammals for a moment and turn our attention instead to intelligent feathered creatures. The corvids, as we have already had occasion to note, are a very well-documented example of intelligent birds. We saw in the chapter 'Bestial biographies' (see pp. 128–37) that crows and ravens are past masters at hiding things. Yet one must first collect food in order to be able to hide it, and what is easier than doing this yourself? Correct – getting others to do it for you. This activity is known as kleptoparasitism. These smart birds, in this instance a species of the corvid family called the California scrub jay (*Aphelocoma californica*), lie in wait in the truest sense of the phrase and observe where other birds have concealed their food. Once they've managed that, they pounce, digging up the food and taking it to a new hiding place. Now of course these canny birds know that they are not the only lazy members of the animal kingdom who would rather pilfer from others, so they are very careful in their choice of hiding place. They keep a close lookout to check that no other animal or bird is watching them when they conceal the stolen food, and avoid making any loud noises in the undergrowth. If they think they're being observed, they grub up their cache again and fly off with it.

Now things get exciting. Not all California scrub jays are as careful as this. Young birds that have not yet begun pilfering from others do not dig up their food if they see another bird observing their cache.[121] Older birds know that, given the chance, they would immediately rob the cache of another animal, and are well aware that others would do the same to them. They mentally put themselves in the position of the bird they are watching and recognize the danger to their own cache. Young birds, by contrast, fail to recognize this. They still don't appreciate that the observer is after their cache, and they are also unable to empathize with the bird watching them and anticipate its future behaviour.

The supreme discipline involved in 'getting inside the head of others' is the insight that others may be harbouring a false belief (see 'The invention of morality'; pp. 147–58). As we have already seen, there's a neat way of testing for a false belief. Unlike the researcher Marthe Kiley-Worthington in France, Michael Tomasello of the Max Planck Institute for Evolutionary Anthropology in Leipzig regards this kind of test as the only valid proof for a theory of mind.

By way of an example, we are going to observe a child by the name of Maxi who has just been given a chocolate bar and who puts it away in a green cupboard. The child then leaves the room and, soon after this, his mother appears. She takes the chocolate bar out of the green cupboard, eats a piece, and then places the bar in a red cupboard. Scientists then asked a group of children, who had seen this scene played out for them with dolls, which cupboard Maxi would look for the chocolate in when he comes back. Four- to five-year-old children give the correct answer that Maxi would look in the green cupboard for the chocolate, since that was where he had left it. But younger children cannot imagine that Maxi does not know what they know and respond that Maxi would look in the red cupboard.

Does this experiment remind you of our pirates in the chapter 'The invention of morality'? Well, if so, you'd be dead right, as Rebecca Saxe[122] expanded her test to include the aspect of guilt and morality. Yet for the pirates too, what it basically boiled down to was whether you could empathize with another person and recognize if they are harbouring a false belief. Tests with animals were run in much the same way, except for the fact that one cannot of course start telling animals stories about pirates or chocolate in cupboards. All the same, tests of this nature that had been adapted to suit animals still saw animals uniformly fail. But was that incontrovertible proof that they do not have a theory of mind? Of course not, since in scientific enquiry one must always allow for the possibility of a false negative result (we look into this further in the chapter 'Experimental errors'; pp. 293–300). Many animals do not behave like humans, meaning that comparative studies of humans and animals are always problematic and not infrequently the subject of heated debate.

For all that, however, comparative studies are extremely popular in behavioural biology, primarily because they are objective – assuming they have been well designed, that is. Yet they can only be deemed to have been properly designed if they are appropriately geared to animals. Experiments in the past clearly weren't, as all the animals tested failed.[123] Yet many behavioural biologists suspected this wasn't the whole story, and so continued to refine the tests.

In contrast to older tests, animals were not now required to perform any confirmatory action but simply to display a clear eye movement. Furthermore, the scene was emotionally charged through aggressive behaviour. Think back to the shell game, and now picture it with buckets instead of cups, with five thick bundles of 100-euro notes. And then there's a bloke who's trying to hide the bundles in such a way that you can't see them and attacking you physically into the bargain. No question that you'd focus your whole attention on the bucket with the bundles underneath it, right? Well, an experiment involving animals would have to be just as attractive and emotionally charged to them, otherwise it might run the risk of yielding a false result, even though the animals possessed the aptitude that was being tested for. You can see the experiments in question for yourself on YouTube.[124] You can very clearly see the eye movements of the test animals there.

The scene was played on television to the test animals and showed the following scenario: the animals saw a person and a person wearing an ape costume. The person in the monkey suit, let's call him King Kong, steals an object from the person and hides it under a box. King Kong then proceeds to shoo the person away and conceals the object under a different box. Of course, that's clever, for in this way King Kong precludes the possibility of the person coming back in and simply looking under the first box and retrieving the object. But now something else happens, and this is the crux of the whole experiment. Unseen by the person, King Kong takes the object from its second hiding place and, taking it with him, leaves the scene. What's of particular interest here is the viewing direction of the ape that is watching this scene play out on television. It of course knows that King Kong has hidden the object in a new place and then taken it with him. Now the person comes back, and what we now want to know is whether the test chimpanzee can read the person's thoughts on the situation. If it had a conception of the person's state of knowledge and knows that the person must have a false belief, then the chimp will look at the first box, under which the person thinks the stone is concealed. Given the fact that the object has now disappeared entirely from view, there is no possibility that it will keep its eyes fixed on the second hiding place. Under these conditions, chimpanzees, bonobos and orangutans all passed the test in 2016.[125] Leaving us humans scratching our heads to try to find another supreme discipline to justify our conviction that we occupy a unique position within the animal kingdom.

Astonishingly, two-year-old children are also capable of passing the test about false belief, as long as the experiments are simplified in a similar way.[126] But here I'll be honest with you and confess that it wasn't a behavioural biologist who devised this clever test procedure. It had been developed ten years previously by a developmental psychologist from London. She wanted to demonstrate that even relatively young children possess a theory of mind with an understanding of 'false belief'. It's truly sobering to think that behavioural biologists spent ten years publishing incorrect findings simply because they hadn't been conducting the right experiments (see 'Experimental errors'; pp. 293–300).

It was precisely this insight that gave me pause for thought, since many animals score just as well in the various tests as children of that age. There is a golden rule in science. If you can explain something simply, you should give up trying to think of a potentially more interesting but also more complicated explanation. If we want to explain Clever Hans' behaviour as 'behaviour reading' (namely, reading body language), then we must credit him with having an extremely good memory and a highly developed capacity for category building. Ultimately, with complete strangers he would need to be able to detect that they were expecting his response at a particular moment. For me, this borders on a miracle. In other words, I don't think it is plausible. Isn't it far more plausible to posit a general system that is applicable in a wide variety of social situations? A theory of mind would be one such system. Here, I simply simulate in my own mind what, in all likelihood, the other person is thinking at that moment. On the basis of my findings from the pointing experiment and my experience with the mousetrap, I am inclined to believe that a great number of animals possess a theory of mind.

This would make many animal behaviours relatively easy to explain. For instance, some years ago a dolphin asked a diver to free it from a fisherman's line in which it had got entangled. It could not expect any help from its own social network, so it swam over to a group of divers and got them to cut the fishing line. Pictures of the diver using his sharp knife to cut the dolphin free went around the world on social networking sites.[127]

A subsequent analysis made it clear that it was a totally wild dolphin that had never before sought out the company of humans nor ever been taken into captivity by them. In my view, such behaviour cannot be explained by anything other than a theory of mind, since we're not dealing here with a cat that had gone feral and had previously been fed by people and that therefore sought out the help of its big human friends in an emergency. Nor was it even a wild animal fed by humans, such as a feral shark that got caught in a strong fishing line and was rescued by members of the same group that was responsible for feeding it.[128] Nor even a young animal, which approached another species community either out of sheer desperation or because it was inexperienced. This was a full-grown dolphin that had evidently come to an independent decision.

Some years ago, my wife, who is a science journalist, wrote a piece on sea cows that had got caught in fishing lines. It was incredible how much resistance the rescuers had to overcome in order to free the poor animals from their plight. That's the normal reaction of a wild animal. Bearing in mind the findings of the mirror test, the decision taken by the dolphin was presumably every bit as conscious as when I go to the dentist with toothache. What's more, dolphins have been shown to support one another[129] and even members of other species[130] in a crisis.

Humpback whales also behaved in a similar way when they tried to save a grey whale calf that had been encircled by killer whales. Fortunately, this astonishing occurrence was filmed by the BBC.[131] In the final analysis, that's just like us carefully restoring a fledgling bird to a nest it has fallen out of. The term used in behavioural biology for this is reciprocal altruism[132] – that is, unselfish behaviour towards others in the expectation of receiving help oneself should one need it. Intriguingly, though, such behaviour really ought to die out, since liars or animals that always exploit others while never themselves investing in the community should from a purely evolutionary point of view have an advantage, and hence would pass on their egotist's genes. The whole thing only works if we assume the existence of a long, perhaps even lifelong memory and presuppose that the animals all know one another. They therefore know who is a helper and who isn't. And logically, those who never help will not themselves receive any assistance.

Yet there are also other observations of animal behaviour in the wild that would be difficult to explain without the presumption of a theory of mind. Thus, a mother elephant was seen to pull a plastic bag out of its calf's mouth before it could swallow it. She kept hold of the bag for a while before getting rid of it surreptitiously.[133] It is hard to explain this behaviour through some basic instinct, for how might such a thing have evolved? After all, plastic bags haven't been around for all that long.

It's still too early at this stage to offer any concluding remarks. Ultimately, we need to examine the workings of our brain and our feelings. Even so, I have in the last few chapters run through the most important criteria and experiments that we scientists use to explain how the faculty of reason functions. I could show you that rodents have the capacity for self-reflection and that fish may recognize themselves in a mirror. We have learned about spiders with personalities and finally discovered that even birds can empathize with others. To be honest, the only thing that I, as a member of the species *Homo sapiens*, can do in addition is to write about these matters.

★

24. Against reason

The invention of minted money in the form of coins around 2,500 years ago helped spur the evolution of human civilization and economic activity. A few years back, the same effect was felt from sequences of binary code on computer hard disks, together with the attendant credit cards and PayPal accounts. We may therefore with some degree of justification assume that we have perfected our dealings with money. But how big a deal is this if small monkeys weighing just 3 kilograms (6½ pounds) apiece are able, after only a few months' practice, to cope with money just as well as we do? The answer is simple: we still have a lot to learn, as the following example shows. The financial crisis and the behaviour of participants in the market shook the world, and the financial misconduct of a few people sent entire economies spiralling into decline. Yet surprisingly the cause of the crisis did not lie, as we would like to imagine, in the greed of a few individuals but rather in patterns of behaviour that we share with other primates. At least, studies on irrational behaviour have indicated this, and it is certainly astonishing to learn that our closest relatives in the animal kingdom make the same mistakes as we do in sophisticated psychological tests. Both we humans and our closest relatives behave just as incorrectly or irrationally as our common ancestors some 30 million years ago. It was at this point that our genetic paths diverged, and unless we happen to have invented the same mistake afresh totally independently of one another, then this pattern of behaviour may go back even further in time. However, to understand this we first need to determine what we actually mean by reason and irrational behaviour.

'Be a good boy and do what your grandmother tells you.' With these words, I would be sent off to spend weekends with my granny as a small boy. By the time I was a bit older, the standard injunction was: 'Just be sensible and don't do anything stupid.' In the interim, several years had elapsed and my brain had continued to develop. My parents were clearly of the opinion that it was time to appeal to my faculty of reason rather than to my sense of obedience. Often, we automatically assume that small children do not have this faculty, and put them in the hands of kindergarten teachers, aunts, grandmothers, and all manner of other responsible adults. They in turn are expected to behave sensibly and to providently guide our little ones with the authority that customarily attaches to adults. We clearly proceed from the assumption that reason only develops in the course of childhood. According to the great German philosopher Immanuel Kant, reason was the chief criterion in distinguishing between humans and animals. For Kant, humans were capable of reasoning, whereas animals were not.

Thinking about such things gives me a queasy feeling in the pit of my stomach. Isn't irrational behaviour at the root of practically every major problem facing mankind worldwide? We continue to burn fossil fuels, despite knowing about their detrimental effect on the climate and the disastrous consequences for future generations that will result from this. We continue to eat vast quantities of red and processed meat, despite the fact that we have known since 2015 that World Health Organization advice is to drastically cut our consumption and so reduce our risk of contracting the third most common form of cancer, namely bowel cancer. We believe in various different gods, and some people even believe themselves instruments of divine will when they blow themselves and others to smithereens with a suicide vest. Understanding the underlying mechanisms of irrational behaviour is therefore of the utmost importance and is currently the subject of any number of research projects in psychology, anthropology, market research and comparative behavioural research.

But now let's attempt, as promised above, to explain our most recent global financial crisis with 30-million-year-old patterns of behaviour. The first thing that springs to mind is a joke I once heard: What do a revolver and Windows95 have in common? Answer: Both are completely harmless until you load them. Anyone who, like me, typed up their thesis on a PC running Windows95 will appreciate the widespread frustration that this joke gives vent to. Yet I'm not concerned here with computer crashes, but rather with the fact that things that supposedly have nothing to do with one another turn out at the core of the matter to be one and the same. Using the methods of comparative behavioural biology, researchers try to identify this core afresh in every new experiment, as in the following example. In the normal run of events, a grape and a dream house do not have much in common. But for a behavioural biologist or a market researcher they are one and the same, since they both awaken desires and the need to acquire what is being presented to us in such an enticing way.

You must surely be familiar with the saying 'A bird in the hand is worth two in the bush'. The following experiment is based on that dictum. Imagine you are a buyer for a medium-sized business concern and that you are aware that the success or failure of this small firm, to which you have been attached for years and which you look upon almost as a family, ultimately depends upon your skill as a buyer. You have the choice of buying from two different distributors. The first one shows you his product, a ball, let's say. To your surprise, after your purchase he gives you not only the ball but a second one as well. What a nice supplier, you think, and perhaps you make a mental note to go back to him next time you're in the market. To your great delight, you have this positive experience over and over again: you pay for one ball and receive two. You feel great. The physiological reason for this is that, as a buyer, your neuronal reward centre kicks in, your hypothalamus releases the substance dopamine, commonly referred to as the happiness hormone, and you float in a buyer's seventh heaven.

But after a while you start to ask yourself if you might not get an even better deal from another supplier. And lo and behold, your boldness is rewarded! The second distributor likewise shows you a ball, but after you have made your purchase, you get not one but two free balls. However, to your enormous disappointment, your second business transaction with this supplier is a far less gratifying encounter: although the supplier once more shows you a ball, after the purchase you don't get anything extra. This whole experience is repeated: you're shown a ball, and one time you end up with three balls and the next time just the one. Seen purely in statistical terms, on average you're getting two balls for one from this supplier as well, but guess which supplier the vast majority of people would plump for? Correct – supplier number one.[134] With him, although there's never any chance of getting three balls, you can count with absolute certainty on getting two, whereas you're regularly disappointed by the second supplier.

Capuchin monkeys operate in exactly the same way.[135] Even though monkeys normally never surrender anything once they've got their paws on it, it was found that they could be taught to carry out genuine financial transactions. After a few months' training, once they'd understood that they could use their toy money to buy food, they started to act in a relatively human way where their purchasing behaviour was concerned. Quite counter to their natural behaviour, which could best be described as 'hand to mouth', they developed a conception of the value of the non-edible means of exchange. And so they pounced on special offers that proclaimed 'Grapes at half price', stole toy money from others, and yes – how could it have been any different? – males began paying out their toy money for sexual favours. The interesting thing about all these studies, which were conducted at various institutions throughout the world, was the possibility of comparing behaviours, which in the absence of the accounting unit money could not otherwise have been compared. Thus, among other things, these experiments played a role in discovering the roots of fairness (see the chapter 'The invention of morality'; pp. 147–58).

But let's come back to the question of our experiment and the financial crisis. We have learned that capuchin monkeys, like us humans, prefer to have a bird in the hand rather than two in the bush. But what happens if we change the experiment ever so slightly and instead of one ball, the distributor presents three balls? Once again, put yourselves in the role of the buyer. You come to the first supplier, and he shows you three balls. You pay your price and get two. Perhaps you'll be a bit annoyed and hope to get all three balls next time. But however many times you deal with this supplier, it's always the same story: he shows you three balls, but you get only two. In frustration, you try your luck with the second supplier. He also shows you three balls, and he really does give you three balls for your money. Yet at your second transaction with him, you're not so lucky, because despite the supplier showing you three balls once more, this time you get only the one. Even so, at least you stand a chance of getting three balls, and you're not about to let that opportunity pass you by, are you? The great majority of people, and also the capuchin monkeys that were being tested   where the balls were represented by grapes – did indeed opt for the second supplier. Quite simply, the first supplier was a disappointment every time. The second supplier was more attractive because with him at least you stood a 50 per cent chance of actually getting three balls. In your mind, you already picture yourself with the three balls, and even though you run a greater risk, since it could transpire that you only end up with one ball, you still opt for the second supplier, with whom the three balls almost seem to be within your grasp.

In purely mathematical terms, there is absolutely no difference between all four examples, since on average you will always have two balls. The form of presentation is the sole deciding factor in which supplier you prefer. Although you will always be certain to get two balls or grapes from the first supplier, it's only in the first experiment that you opt for him. If our reason was guided even only partially by logic, we would opt for the first supplier in the second experiment as well, since when all's said and done, we always reliably get two balls from him.

But why do we go for the second supplier in the second experiment, given that we run the risk with him of only getting a single ball? Researchers call this attitude 'loss aversion'. If we possess something, or in this case believe that we are almost in possession of it, we're reluctant to relinquish it. In our case, the researchers were at pains to devise an experiment in which it ultimately didn't matter in the slightest what your preference was. Accordingly, a wholly rational person would sometimes opt for the first supplier and sometimes the second. But in reality, it is not a matter of complete indifference what we opt for, because if as a result of loss aversion we take a risk, that risk may be far greater than our potential gain.

And it's precisely such behaviour that brought about the 2008 financial crisis in the United States. On the one hand, fantasies, slick marketing and wishful thinking inflated the price of various financial products, and on the other stockbrokers couldn't shift these profitable and much loved securities fast enough. Yet it wasn't only stockbrokers who were affected by this phenomenon. When the first effects of the crisis began to be felt in the labour market, thousands upon thousands of house owners found themselves unable to pay off their mortgages. They clung on to their houses for far too long. The aversion towards losing their 'own' four walls was so great that most of them waited until the game was up and the entire real estate market had totally collapsed. The houses came under the hammer for a fraction of their original value, and their former owners found themselves without a roof over their heads but still with a considerable chunk of debt owing to the banks. They in turn derived no benefit from this huge number of borrowers, since they were no longer in receipt of the regular mortgage repayments, and so a vicious circle was set in train.

This knowledge is part and parcel of every sales force's or agent's training course, however banal. You deliberately set out to convince people that they really need a product, and you give them the sense that they are already in possession of their object of desire. If the salesperson succeeds in instilling this state of mind, then as a buyer you are virtually powerless, and the prospect of letting the object of desire slip through your fingers is unbearable. It is a fortunate person who, at that moment, suddenly thinks it might be better to simply go home and shop around in their own good time for a better offer on the Internet. But with a bit of luck, the pain of the supposed loss will dissipate at home, and you can spare yourself the search anyway.

As a creature endowed with the power of reason, you will have cracked another enigma of the animal world and you can finally break free of these 30-million-year-old fears of loss. You won't fall for salespeople's tricks ever again, you'll take up a diet containing little meat and you'll renounce the use of fossil fuels.

Let's turn our attention to another irrational piece of behaviour, the so-called decoy (or bait) effect. This is a well-known ploy in marketing. If, for example, a firm wants to sell a mobile phone and gives you two alternatives to choose from, then you will base your decision on some criterion that's particularly important to you. It might be the question of affordability. Mobile A costs £350 and has 64GB of storage, whereas mobile B costs £250 but has only 16GB. Presumably you'd go for the cheaper version, since when all's said and done, 16 gigabytes still gives you plenty of storage space, and if need be you could always upgrade your memory card. Yet the company's profit margin will be larger if their marketing strategists can talk you into buying the more expensive phone. The few pounds extra it costs the firm to provide more storage space is negligible in terms of their investment. Because they know how the decoy effect works, the salespeople now throw another offer into the mix. You're now faced with three options.

Offer of three mobile phones with different storage capacity

	Mobile A	Mobile B	Mobile C
Price	£350	£250	£400
Storage capacity	64GB	16GB	32GB

In all likelihood, offer A doesn't look nearly as unattractive to you now, and indeed many buyers do opt for A in these circumstances, and the marketing gurus coin it.

Incidentally, Tanya Latty and Madeleine Beekman, biologists at the University of Sydney, believe that this form of economically irrational behaviour is both deep-seated and sensible in evolutionary terms. Among other things, they base their hypothesis on the observation that other creatures such as bees and birds make similar mistakes, and even go one better: they tested the slime mould species *Physarum polycephalum* for the decoy effect,[136] and it returned a positive result. It seems incredible but it's true all the same: this slime mould would also have gone for mobile A – of course, only on condition that it was edible. The basic principle was the same, though. The researchers provided two alternative food sources, both of which had advantages and disadvantages. The slime mould could not decide and simply grew in both directions. But then a third yet unattractive option was thrown into the mix, and lo and behold! – the slime mould did not come to a rational decision, which one might have thought it, as a single-celled organism, would have done purely mechanically. No, instead it inserted a decision-making process and, just like us, weighed up the various options and ultimately went for one of the two original alternatives. In comparison with the third unattractive option, one of the two original options simply seemed more attractive. Naturally, it wasn't exactly a compliment to the quality of human decision-making to discover that we're ultimately no better at it than a slime mould.

But it gets even better: another experiment succeeded in showing that this simple, single-celled organism even had a memory, and what's more without possessing a single nerve cell. It simply marked its path chemically, in much the same way as ants do with their trails.[137] In their experiment, the biologists also took inspiration from spatial tests carried out in robotics research. Thus, robots that have been designed to mow lawns need to orient themselves within a spatial environment that they have no conception of. Just like our slime mould.

The fungi researchers take the view that our irrational behaviour isn't at all irrational, at least not from the perspective of evolutionary development, for in most cases it makes perfect sense to come to a decision on the basis of taking a long, hard look at the various options and weighing them up. So, for our fungus, it might not be a sensible strategy to go for two food sources at the same time, while animals that cannot simply split themselves in the middle must necessarily decide one way or another anyhow. But if, when faced with complex problems involving three options, a rational decision can only be managed by applying higher cognitive processes, then a decision on the basis of simple comparison is a logical and successful approach over the course of evolution. Finally, just a brief amendment of my terminology here: scientists recently decided not to classify slime moulds as fungi any longer, but as genuine single-celled animals.

What's more, comparative irrational decision-making also has its advantages in a social context of cognitively more highly developed animals. The Yale professor Laurie Santos, who was responsible for devising the aforementioned experiments concerning the financial crisis, is of the opinion that loss aversion is an effective means of not losing sight of social services, such as coat-grooming among primates. The animals constantly compare to find out whether they are doing more grooming than they receive themselves. The absolute measure of the time spent on this task is wholly irrelevant; all that matters to them is a fair distribution. In this way, the animals ensure with a fair degree of success that they are not short-changed. Ultimately, then, this approach is not a mistake but simply an effective way of getting to grips reasonably effectively and equitably with something extremely complex like the exchange of social courtesies.

What we are left with is the insight that our irrational behaviour has thoroughly rational causes. All the same, this does not absolve us of the responsibility of trying to solve the global problems resulting from our irrational behaviour. The only question is whether we place this responsibility in the hands of each and every individual and hope that they can emerge from beneath their evolutionary shadow. Yet perhaps politicians and other movers and shakers in our society also ought to anticipate the irrational behaviour of individuals within the mass of humanity and guide people using instruments of marketing with the same adroitness that sales strategists currently employ when targeting consumers? For instance, Gerd Gigerenzer, director of the Harding Center for Risk Literacy at the Max Planck Institute for Human Development in Berlin, takes the view that it is absolutely legitimate to manipulate people if it is in their best interests.[138] This thought brings us full circle, since after all we manipulate and guide our children, who have not yet developed the capacity for rational thought, without any moral scruples whatsoever.

25. The thinking apparatus

During my studies, I spent several years working night duty on a neurosurgical intensive care ward of the University Clinic in Kiel. One day a young man was brought to our ward who had just undergone an operation lasting twelve hours. He had put a gun to his head and pulled the trigger. Although the emergency operation prevented further bleeding to his brain occurring, he had irreversibly blown away a third of his brain's frontal lobe. This meant that, directly behind his forehead, there was nothing left of his brain. He not only survived the operation, but three days after his surgery I was able to communicate with him. Of course, he was still on a ventilator, but so were almost all the patients on our ward, and you soon learn how to cope with that. Under these circumstances, one cannot exactly start engaging in philosophical debate, but you can ascertain whether someone is feeling pain, or has a bad taste in their mouth, or is hungry or thirsty, and also whether a person is sufficiently clear-headed for you to be able to safely untie their hands from the bed frame.

The young man was surprisingly fit and presumably had not even noticed that large parts of his brain were missing. But how could that possibly be? The so-called prefrontal cortex is responsible for brain functions such as problem solving, forward planning and targeted action, as well as personality traits. In other words, all things that you can do without if need be. I assume the young man would only have realized some weeks later that he was doing things differently from the way he had done them in the past. Generally speaking, he would have felt less desire to do anything at all, and presumably he would also have been unable to summon up the strength and the inner motivation to make another attempt at suicide, since that kind of decision is also made in the prefrontal cortex. It is something of an irony of fate that he shot away the very part of his brain that drove him to attempt to take his own life in the first place.

Our brain is therefore divided into many different areas. While these areas are sometimes tightly and sometimes only loosely inter-linked with one another, they do each perform very distinct tasks. If part of the brain ever gets damaged, then we find it extremely difficult, or even impossible, to regain certain faculties. We have learned that small areas of the brain such as the right temporo-parietal junction (rTPJ) are responsible for certain functions and so form partial aspects of our personality. Perhaps you can recall the question of magnetic blocking of the rTPJ area (see 'The invention of morality'; pp. 147–58), which renders those affected incapable of reaching morally sound decisions. This brain area is only the size of a pea and emphasizes how tremendously important every single part of our brain is. Sure, we can do without a good deal of brain mass and still go on living without any problem, but we are no longer the person we once were. Certain functions or aspects of our personality have vanished – a small part of us has effectively died.

And yet, as a result of such experiments or accidents, we have a very good understanding of how our brains work. We know, for example, that there is a form of brain damage that makes you incapable of recognizing your own mother, even though she's stand-ing right in front of you. As soon as your mother speaks, however, you recognize her voice and so acknowledge her as your mother. A similar thing can happen to you, in such a state, when you look in a mirror. Although you'd still pass the mirror test, you don't recognize yourself. Such extreme examples,[139] which are clearly associated with injuries to particular parts of the brain, emphasize the high degree to which our cerebrum is specialized.

In addition, it is only a fraction of our thoughts that we think consciously or formulate in language. The most important decisions in our lives – namely those we find most difficult and that we first need to sleep on – are not ones that we make consciously. These important thoughts are often so complex that we have no linguistic equivalent for them, and so the notion that thinking is only possible through language must be regarded as obsolete.

Watch out, though – things start getting uncomfortable here. Normally, we think we are thinking our conscious thoughts 'live' and believe we are completely free agents in what we think. Right now, for instance, I'm mulling over this sentence in this book I'm writing, but I'm totally at liberty to decide to quickly click on Amazon and order myself a T-shirt or get myself a new Blu-ray from my online video library. Not a bit of it, though! The renowned neuroscientist Michael Gazzaniga has spoken of human free will as being a fairy tale that our own brain recounts to us.[140]

You don't believe it? Just imagine you're sitting in an apparatus that can measure your brain activity. All you're required to do is decide whether to press a button in your left or your right hand. It's left entirely up to you which button you want to push when, there are no stipulations. It's all a matter of your free will. The only requirement is that you must immediately, namely the instant you have decided, press one or the other button, really press it firmly. Thought and action should occur simultaneously in the twinkling of an eye. So how much time do you reckon it took your subconscious to decide to press the left button? One millisecond or ten, perhaps even half a second? No, from the data on his or her computer screen, the experimenter can predict fully ten seconds beforehand which button you are going to push.[141] In the case of more complex tasks like simple arithmetical problems, the experimenter can predict the result at least four seconds ahead of the conscious decision.[142] This is really alarming, agreed? To help us grasp all this, and to enable us to judge what the difference is between a human and an animal brain, we need first to settle on some basic ground rules.

To a behavioural biologist, the brain is somewhat akin to a black box; we don't necessarily have to know what's in the box, as the main thing we're interested in is behaviour and that can be observed fine without knowing what's going on in the brain. Nevertheless, you can't engage in any discussion about intelligence in animals nowadays and not consider aspects of brain anatomy, so I need to deal with a few basics here first.

First, we need to understand that the invention of the nerve cell – that is, the structure through which we think – is ancient. At the very moment when single-celled animals began to lead a social existence, thereby giving rise to multicellular animals, the problem of stimulus transmission arose. A few million years later, the first nerve cells came into being and strictly speaking not much has changed in their basic construction ever since. Today, we distinguish between two groups of nerves. One group has something resembling a pancake of fat wrapped around them, whereas the other group does not.

We call one group of animals protostomes (literally 'first mouth', from the fact that the oral end of the animal develops from the first developmental opening, the blastopore); these include worms, insects, mussels, snails, spiders and crabs, and they have no 'pancake'. The second group are known as deuterostomes ('second mouth', in which the oral end of the animal develops from the second opening on its dorsal surface). These include the sea urchin but also all vertebrates, such as fish, reptiles, birds and mammals, and hence also us humans. One key characteristic of this latter group is that they began at an early stage of their evolution to eat with their anus. Yes, you read that right – they ingest their food not with their original mouth but with their rear end. But let's leave aside these niceties of our development and concentrate again on our nerves. The advantage of the aforementioned pancake-like cells, the so-called myelin sheath, is that they isolate the nerve axons. These are the projection of individual nerve cells. They can be over a metre long and are responsible for transmitting stimuli. Because our nerve cells also transmit electrical impulses, this isolation is of great benefit. It not only saves energy but is also faster.

For instance, if a giraffe were to injure its foot, then it would take half a second for it to draw its foot back if it did not have this isolating sheath. In actual fact, it takes just a tenth of a second. This time span can make all the difference between whether we injure ourselves seriously or not.

A further major step in evolution was the emergence of gang-lions. These are clusters of nerve cells in which the first compu-tation processes occurred. These processes really are very similar to the calculations made by a computer, since our nerves also communicate digitally with one another, in other words using binary code of zeros and ones. A long sequence of ones one after the other signifies a high intensity of stimulus. Unlike in computers, in which the individual transistors are each capable of processing only one signal, nerve cells can simultaneously receive numerous impulses from other nerve cells. The truly magnificent invention, however, is the synapse, namely the connecting points between one cell and another. Although the conduction of stimuli in our nerves takes place electrically, the transmission between the individual nerve cells is a chemical process. For this reason, there exists at the end of every nerve cell a synapse, which converts the electrical signal into a chemical one. The electrical impulse triggers the distribution of so-called neurotransmitters, which are received on the opposite side, the so-called postsynapse. If these messenger substances are detected, a new electrical signal is generated.

You may well have read or heard that the flavour enhancer monosodium glutamate is bad for your health, and so avoid eating packet soups or other prepared foods. But in fact that's not correct, since glutamate is one of the body's own neurotransmitters and is decidedly not poisonous or unhealthy. On the contrary, without it we would be unable to taste a thing. The fact is that packet soups and other prepared meals actually taste of nothing. But if you add monosodium glutamate to this food, which arguably doesn't even warrant the name, our postsynapses are bombarded with neurotransmitters and think mmm, this is tasty! Many drugs act in a very similar fashion; our postsynapses are fooled into thinking there's something there, which in truth doesn't really exist. We'll go into this in a bit more detail in a moment, in part VI, 'Sentimentality' (pp. 269–82).

Put very simply, there are two different types of synapse; some transmit an incoming signal onwards, while others do exactly the opposite. The latter type generate a negative potential, which can cancel a positive signal from another synapse. In a lecture I recently heard, a neurologist spoke in terms of a war between synapses. Each individual nerve cell receives countless positive and negative impulses and has to process these. Ultimately, the sheer volume determines how active the receptor nerve cell is. Only if the positive impulses predominate will it transmit the signal onwards. This identical function is involved in bending or stretching your knee or in reading and thinking. If you read the word 'leading', the first thing that happens is that all the nerve cells in your brain that are wired for the letter 'L' start firing. Then the same thing happens for the letter 'E' and so on. All other letters, each of which also has a set of interconnected nerve cells, do not communicate because their negative potentials are predominant. It's hard to believe but it really is that simple. At this level, there is scarcely any difference between ourselves and other animals. Also, the five areas of the brain (the telencephalon, the diencephalon, the mesencephalon, the metencephalon and the myelencephalon) have comparable functions in all vertebrates.

In nature, almost every living organism begins as a vesicle, from the surface of which a germ layer develops, which in turn may give rise to an organ or tissue. In the case of our brain, a structure developed with many layers. You may have heard that the brain contains two kinds of tissue, known as grey and white matter. Grey matter is the nerve cells, while white matter constitutes the pathways (axons) connecting the individual nerve cells. Thought processes are therefore carried out exclusively within the grey matter of the surface layer, since this is where the processing of the individual action potentials takes place. Unfortunately, a surface layer can only ever be as large as what surrounds it. To solve this problem, in all mammals the cortex – that is, the superficial layer of the telencephalon – is flattened into folded pleats. It was once thought that the extent of this convolution was an indicator of brain power.

Some fifty years ago, however, dolphins were found to have a far more folded and convoluted cortex than we humans. Subsequently, it was suggested that nerve density – that is, the number of nerves per surface – was the key factor, until it was discovered that mice have a higher density than us. Then, around twenty years ago, the media were abuzz with the discovery of mirror neurons. All of a sudden, here was something that existed only in humans and some other ape species. Mirror neurons, it was claimed, had the amazing capacity of simulating in us the thoughts and feelings of others. This is, as we have seen above, of enormous advantage to animals that live socially. In this way, I can empathize with another individual and anticipate his or her behaviour. This gives me the potential to adjust my own behaviour in accordance, thus enabling me to sympathize with others, or to deceive or manipulate them. It was even speculated that mirror cells might be responsible for our complex social life and that we humans ultimately owed our enormous social skill, indeed even our whole make-up as humans, to them.

Perhaps the most famous example of the function of mirror neurons is the act of yawning. When we see a close social partner yawn, we also feel tired and start yawning ourselves out of solidarity. In this way, a social community ensures that exhausted members are also taken notice of, and that we as a group do not exceed the effective capacity of individuals. In other words, this is a profoundly social and considerate behaviour.

But if we look carefully, it is clear that it is not only apes that yawn out of solidarity, and so it came as no surprise to me to learn that communal yawning can also be found among wolves[143] and even budgerigars.[144] Dogs, our favourite quadrupeds, yawn in concert with us.[145] One of the discoverers of mirror neurons, Christian Keysers, has advocated conducting experiments with rodents, because they are so close to us in their emotional behaviour.[146] To date there have been almost 30,000 publications on mirror neurons, yet we still do not fully know what they do, or all of the animals that possess them.[147] Accordingly, there is now an increasing tendency to talk about 'shared neural activations' – namely, a diffusely distributed neural network that operates as a simulation mechanism within us.

A less contentious type of cell are the so-called spindle neurons, which were rechristened 'von Economo neurons' after the scientist who discovered them. These cells solve a problem present only in large cortexes. In contrast to the majority of nerve cells, spindle neurons have only one entry point and one exit point. As such, they are not involved in the actual thought process. Rather, their task is to connect distant parts of the brain and so create circuits that would otherwise not exist. This appears to be particularly crucial regarding social problems,[148] since these require different areas of the brain to interact intensively with one another. For a long time, it was thought that these cells were only present in humans and great apes.[149] However, further investigations have shown that this cell type also exists in whales and dolphins[150] as well as in elephants.[151] But to infer from this that a mouse has a lesser capacity to think or interact socially is dangerous, for it may be the case that this type of cell does not exist in mice simply because there are no great distances to be overcome in their brains.

Yet it is not only among mammals that intelligent brains have evolved. A bird's brain, for instance, does not have a convoluted cortex. If we were to employ the yardstick of our heavily convoluted cortex, we would have to consider birds as rather stupid, since their brain does not have any surface enlargement and hence not much space to think. To counter this assumption, a team of scientists recently entitled their paper 'Cognition without Cortex'.[152] In actual fact, birds have taken a different evolutionary path. Their brain is not some hollow bubble where thinking only takes place across a limited surface area. In contrast to our brains – in other words, the mammalian brain – the entire cortex thinks as one single mass. Presumably this helps save a great deal of weight. In any event, it explains why a brain weighing only a few grams can be so relatively efficient and score incredibly well in numerous experiments.

Very recently, an article was published that posited a comparison between the capacity for language acquisition in small children and the way in which birds learn to sing.[153] It was hoped that this approach might yield a better understanding of how language evolves, in comparing systems that were equivalent but which had developed entirely independently of one another. Anyone struggling to believe this might find the following observation helpful: we are all familiar with the devastating effect of alcohol on our ability to speak. Scientists have thus far been unable to fathom the neurobiochemical mechanism underlying this, but it is hoped that studies of finches under the influence of alcohol may throw some light on it. For, just like us humans when we have overindulged, the birds begin to slur their songs.[154] Which actually brings us straight on to our next topic: do animals take drugs, and do they treat themselves with medicines?

But before we consider this fascinating topic, I should acquaint you with an important trend in brain research. As you know, the principal reason for conducting brain research on animals is to better understand the workings of the human brain. Ultimately, the sale of drugs to combat not only Alzheimer's disease, Parkinson's disease, multiple sclerosis, epilepsy, headaches and migraine, but also depression and schizophrenia, is potentially big business. Yet thanks to the considerable anatomical and physiological similarity of their brains to ours, fish also became research subjects some years ago.[155] The joke's now firmly on any angler who still believes the old fishermen's tale that fish can't feel pain and are incapable of suffering.

26. Shamans

The term 'shaman' came originally from Siberia, and denoted wise men who, with the aid of their knowledge of the pharmaceutical use of plants, acted as healers within their community or alternatively put themselves into a state of narcosis in order to get in contact with the spirit world. A number of Siberian brown bears must have felt much the same way when, on the Kamchatka Peninsula in the Russian Far East, they found that they could get wonderfully high by sniffing empty kerosene drums. The German newspaper

Die Welt even ran a headline in 2013 that read 'Kamchatka's Drug-Bears Hooked on Kerosene';[156] the article highlighted an ecological problem inadvertently caused by agencies that had come to protect the environment there. Dotted all around the forests of the peninsula, they had left unsecured a large number of kerosene drums, put there as a fuel reservoir for the helicopters they used to get around in. Bears are inquisitive animals: visitors to Yellowstone National Park in the USA are warned to store their food in well-sealed containers and bear-proof lockers. And so it didn't take long for the Siberian bears to discover that the drums contained something good and to work out how to use them.

Even so, there is something rather funny about this story. Grizzly bears (*Ursus arctos horribilis*) are a subspecies of the brown bear (*Ursus arctos*) and lead a solitary lifestyle, as do their close relatives on Kamchatka, the two populations being separated only by the Bering Strait. They are not nomadic but have a clearly demarcated territory, which is even marked by boundary trees. The animals seek out mountain ridges with good visibility and free-standing trees, which they mark at roughly chest height with claw scratches and bite marks. Male rivals can immediately tell the size and strength of the territory's owner from the height of the markings, while females sniff them to assess the male bear's sexual prowess.[157]

Bearing in mind their penchant for limited territories, it makes no sense that the bears on Kamchatka, as described in the report in *Die Welt*, should have travelled long distances to binge on the kerosene stores and then sleep off their hangovers. In my view, the probability of being caught staggering around a strange bear's territory and getting a savage mauling is really very high. My scepticism duly led me to do some digging in the scientific literature. And indeed, I found only a single publication in which this behaviour received a mention, and that over a mere five lines. The researchers spoke of the preference for sniffing kerosene being dependent on the age and sex of the bear. They also cited a mother and cub pair that, in contrast to the other animals observed, returned to a kerosene dump on a daily basis. After the mother had died, the reason for which was unspecified, the young bear continued to display the same behaviour.[158] So, on the basis of these isolated cases, it seemed a bit excessive to run a piece about an entire bear population on Kamchatka that was hooked on kerosene. Yet what is the real story behind the mystery of animals and narcotics?

Under the influence

At the end of my studies in Kiel, my wife and I planned to help crew a catamaran sailing from Panama City to Florida. It was supposed to be a great adventure, as the two of us really love the sea, and we could already picture ourselves lounging on idyllic Caribbean beaches. Having arrived at our hotel in Panama City, we flung our bags in the corner and headed straight out to the yacht marina on the Panama Canal. It promised to be a lovely stroll along the coastal promenade. Sadly, our excursion ended in a police station, with a friendly policeman marking out a couple of circles for us on a city map. Best avoid these areas if you're a tourist, he commented tersely. And we'd just walked straight through one of these circles. That evening, we sat with our room neighbours on the balcony of an old hotel built in the colonial style, looking down at the Plaza de la Independencia. The air was warm and humid; the duty-free bottle of Jack Daniel's tasted wonderful and helped me put out of my mind the image of the four armed youths, the knife held to my stomach and my dread for my wife's safety. The whiskey tasted so good that evening that I resolved never to touch the hard stuff again thereafter. This was a conscious decision that I reached as a result of this unfortunate experience. But why should alcohol be so attractive and so feared, and are there animals with an affinity for spirituous liquors?

Fortunately, we humans are not the only ones to indulge in this vice. A few years ago, a report did the rounds indicating that fruit flies (*Drosophila* sp.) that did not get enough sex took solace in alcohol.[159] The study behind the report investigated the molecular mechanisms behind the effect of alcohol on how addiction arises. The flies were found to have so-called neuropeptide Y, a messenger substance that is also produced by our nerve cells. The researchers were interested in this molecule because there is a suspected link between stress and both obesity[160] and alcoholism.[161]

Naturally, the media had a field day with this theme of sexual frustration and alcohol consumption. The analogy with us humans was just too perfect, since ultimately it turns out we aren't the only ones to drown our frustration in drink. But this wasn't at all the intention of the report's authors, whose sole concern was the molecular mechanism, and so two years later, a specialist journal for fruit-fly research (yes, there really is one!) put the record straight by announcing that flies like alcohol for quite different reasons than we do.[162]

Alcohol is almost as high-yielding in energy as sugar, and flies rely on this outstanding source of sustenance. This is especially the case for flies that have not yet been able to indulge in their urge to procreate. These animals were simply not impressive enough to the females, being too small and weedy. In the final analysis, they had not yet fed themselves up to the requisite size of an attractive male. Unlike us, flies can't just chomp heartily into an apple. They have to wait until the apple falls from the tree and the cellulose in the fruit's cell walls is broken down by microorganisms. Even though a crisp, fresh apple really appeals to us, we can't break down the cellulose that makes it nice and crunchy either. That task is taken on by bacteria in our large intestine. Either way, rotting fruit doesn't hold much attraction for us, but to fruit flies it's a delicacy, since it is only when the fruit takes the form of this slimy, deliquescing mass that they can suck up the food with their proboscises.

A side effect of this process of decay is the production of alcohol. Because flies also don't want to be buzzing around on a permanent high, they have evolved enzymes that help them to break down alcohol and turn it into energy. In this respect, it's quite right that flies don't try to drown their sorrows in alcohol but instead nourish themselves from it, and thereby grow larger and more attractive to their females. As a direct consequence of this they are no longer sexually frustrated and hence need less alcohol to grow.

But leaving this aside, there are still many animals that, just like us, really enjoy getting hammered. An old BBC documentary[163] shows pigs, baboons, squirrels, giraffes, elephants and other animals congregating beneath a marula tree in Africa – also known as the elephant tree – in order to partake of its fermenting windfall fruit.

Several years ago, when hardly anyone in Germany had heard of Amarula Cream, I was sitting on a plane on the way back from South Africa to Germany. This liqueur, made from marula fruits, tasted truly divine and made the flight back more bearable. Who knows, maybe the distillery in South Africa that started producing it did so after someone saw the BBC film. In any event, it was thanks to financial support from the distillery that the 'Amarula Elephant Research Programme'[164] was established at the University of KwaZulu-Natal, which has helped the cause of elephant conservation.

But here's an odd thing: scientists have worked out that elephants at least cannot possibly get drunk on the fruits of the marula tree. Even assuming the highest possible concentration of alcohol in the fruits, an elephant would need to consume more than 5,000 of them to get drunk. Yet the maximum daily intake of marula fruit by elephants is only around 700.[165]

An old friend of mine, Solvin Zankl, is a photographer for the magazine *Geo*. He told me that there's a strong code of ethics among nature photographers these days, and that no editor would ever take any pictures from him that looked like they had been staged. There was a time, though, when even the greatest photographers and documentary film-makers had been all too keen on faking scenes. Flipper, the idol of my childhood, wasn't one dolphin, but several. As a result of poor conditions in captivity, the animals did not survive for long and were simply replaced on a regular basis without the audience noticing. A favourite trick in documentaries was to tether a goat as bait for a predator, and I wouldn't be at all surprised to learn that the drunken animals in the old BBC documentary had been given a drop or two of the hard stuff. But the fact remains that those images of drunk animals left their mark on a whole generation of viewers. The idea that we weren't alone in our vice was just too seductive. But what's the real story, or is it all just a sham on the part of the media?

In India, there have already been a number of accidents in which drunken elephants have injured either themselves[166] or people.[167] In all cases, though, it was not a natural source of alcohol that was to blame. That would presumably have been too much like hard work for these clever pachyderms, so they stuck instead to rice beer, a popular and cheap beverage in the remote villages in the north-east of the country. Wallabies in Tasmania displayed a similar opportunism in feasting on plantations growing opium poppies.[168] Let's hope for the marsupials' sake that Australia continues to be a market leader in the production of medical opiates.

In North America, it's farmers who most frequently come across drunken animals. Both cows and horses seem to be mad for locoweed, a legume of the Papilionaceae family. Its name derives from the Spanish word *loco*, meaning 'crazy'. Various myths have long circulated regarding these grazing animals and their predilection for narcotics. Yet a scientific investigation came to the conclusion that locoweed was often the only green plant available on arid farmland and the animals naturally preferred it to desiccated grass. If they had the choice of fresh pasturage, they'd no longer give locoweed the time of day and at most would only ingest it by mistake while grazing grass.[169]

If you browse the Internet, you could easily go on adding to the list of drug-addicted animals for as long as you liked. There are reindeer that eat fly agaric fungi, cats that adore catnip, alcoholic bees that are banned from flying and dolphins that pass around highly poisonous blowfish like a joint.

No question that a wealth of anecdotes and observations exists out there about animals that have got drunk, or stoned on other drugs. But evolution knows no mercy and if an animal is high, it is a sitting duck for predators or other enemies. The burning question is therefore whether animals willingly expose themselves to the influence of mind-altering substances. Do they consciously decide to eat fermented fruit or sniff kerosene drums? Or do these things happen by accident? You could argue that the animals in question would not repeat the experience if it was unpleasant or dangerous for them. Yet though that sounds logical, it needn't necessarily be the case. Thus, for example, a dog would not hesitate to run through a thicket of brambles in pursuit of a scent it had picked up. Even if it injured itself on the thorns, it would do it again; all it takes is for the stimulus to be sufficiently strong. It must be exactly the same for a bear that has learned that people's leftovers are often delectable titbits. Maybe that was why they started sniffing the fuel drums. When all's said and done, kerosene is an oil, and oils are food.

If I were called upon to decide here and now whether animals take drugs voluntarily in order to intoxicate themselves, I would say no. Although it would really please me and I would be happy to grant the animals this pleasure, we have to concede that at present there is no substantive evidence arguing in favour of voluntary drug consumption. On the contrary, a large number of observations lay themselves open to a different and more plausible interpretation. This is especially true of instances of repeated alcohol consumption by animals.

There is even a theory claiming that practically all fruit eaters have a preference for alcohol. To grasp this, you first need to understand the following context: the great majority of our planet is subject to seasonal changes to which all living organisms are attuned. Particularly for animals, surviving the cold season represents a major challenge. Since time immemorial, sweet fruit at the end of the growing season has guaranteed a high chance of survival over the coming winter, since the sugar in the fruit is an excellent energy provider. Half of it is in the form of glucose, which cannot be stored and is therefore used in the course of a day, and the other half is fructose, which is instantly converted to fat in the liver. Fat is ideal for storing, and if we put it on in the form of subcutaneous fat tissue, it can even provide insulation from the cold.

And so at the end of the summer season, our bodies get on the one hand the correct raw materials for laying down fat and on the other a sweet signal reminding them that they should get on and do precisely that. Nutritional scientists even see in this situation a connection with excessive weight. Throughout the past millions of years, there was only a single opportunity each year to eat sweet things, namely in the form of fruit. Our bodies are programmed to extract as much fat as possible from the fructose and to store it for the ensuing cold season. In the Western world, however, sweet things are available all year round, and so our bodies are constantly receiving the signal to turn the sugars into fats and store them. This context also explains why so many Americans become obese despite a diet of reduced fat (but sadly also sweetened) foodstuffs. Fruit therefore plays a key role as a source of nutrition and an energy store.

Yet sugar in fruit is not only attractive to animals, and every fruit that drops from the tree – and ideally splits open into the bargain – is immediately colonized and decomposed by microorganisms. However, just a few millimetres below the surface, no oxygen is available, and the process of alcoholic fermentation begins. Alcohol is a highly volatile substance, though, and all fruit eaters have a good nose for it. Animals respond to the smell of ferment-ing fruit in the same way we do when we swirl around the wine in our glass to enjoy its 'nose'. It makes no odds whether I am a 700-kilo (1,550-pound) grizzly bear or a little songbird. Alcohol exerts a magical allure on everyone, as it means sugar, and sugar means fat. Fat as insulation, fat as a reserve of energy – something that will enable you to survive the coming winter. However, lest I be accused here of stealing another's thunder, I must admit that these ideas are not mine, but those of Robert Dudley,[170] Professor of Integrative Biology at the University of California at Berkeley.

That said, there are unquestionably instances of addiction in the animal kingdom. For instance, the acacia tree lives in sym-biosis with ants, which protect it from pests and which in return are provided with nutrients by the tree. Nonetheless, the relation-ship isn't actually as unambiguous as that makes it sound, as the acacia begins by making the ants dependent upon it. It does this by producing a toxin that inhibits the ants' digestive enzyme, invertase. Without this enzyme, the ants are unable to digest sugar. The only available food source for them is the nectar of the acacia, but this of course contains the toxin. And so in this way the tree maintains its own garrison of addicted defenders.[171]

*

Chickenpox parties and other forms of medicine

If you have ever lived for any time with a domestic cat or dog, you'll know that they eat grass now and then. The reasons for this behaviour are various: the high proportion of roughage in grass stimulates their digestion, and taken in large quantities can cause them to vomit; it can also help them combat pathogens. This does not signal that the animals are sick; it's simply part of their normal spectrum of behaviour. Young dogs eat more grass than adults, but only adult dogs vomit.[172]

Unlike the question of drug consumption, self-medication and prophylaxis in animals is a far less contentious issue. There are countless examples of animals treating themselves with the aid of plants or even other animals, or of using them to ward off illnesses.

Eating grass helps digestion because the roughage in it provides food for the microorganisms living in the dog's intestines. Yet it also appears to help combat undesirable germs. A study aimed at improving dog nutrition demonstrated, for instance, that certain plant components reduced unwanted bacteria in the intestines of dogs.[173] This tallies with observations of raccoon dogs.[174] These studies showed that sick raccoon dogs had more grass in their digestive tract than healthy ones. Perhaps you're asking yourself how scientists could possibly know about the stomach contents of healthy and unhealthy raccoon dogs? These appealing animals, which get their name from the fact that they really do look like a cross between a raccoon and a dog, are not among the indigenous fauna of Europe, but were originally released as game animals. Accordingly, it has not been hard for researchers to get hold of dead raccoon dogs. In Germany alone, the hunting bag[175] from 2003 to 2013 totalled 250,000 of these animals,[176] equal to the human population of a sizeable town.

Without a doubt, eating grass is not some arbitrary behaviour on the part of animals; they do it for a good reason. But what's going on in their heads when they do so? For most people, this is a classic example of animal instinct. As you will already have had cause to note from the earlier section on culture, I take a very critical stance regarding the concept of instinct. Not to beat about the bush, I find the term meaningless, since it explains absolutely nothing. Our ancestors thought that thunder was sent by the god of thunder. But at some stage we began to understand the workings of the atmosphere and as a result we now know what the real root causes of thunder are. It's much the same story with the term 'instinct': it describes an observation that could not previously be explained in any other way. For Charles Darwin, for example, instinct was innate behaviour that had developed over the course of evolution as a result of natural selection. Meanwhile, the Dutch ethologist and Nobel Prize Laureate Niko Tinbergen differentiated more precisely between inherited patterns of behaviour (ultimate causes) and immediate triggering factors (proximate causes).

Yet despite extensive research, the population perception of this subject has changed very little. Instinctive behaviour is contrasted with rational behaviour. This way of looking at things underpins the perceived clear distinction between the rational human being and all other animals, which are deemed to be prey to their instincts. But over the course of this book we have clearly seen that this division really ought not to be upheld any longer. Even so, let's investigate briefly how it might feel or what a person might be thinking if they behaved 'instinctively'?

Just imagine you're thirsty. The osmoreceptors in your hypothalamus, an important part of the brain, open mechanical ion channels and generate a tiny electrical short circuit. This short circuit speeds as an action potential to the neurohypophysis, which consequently produces the antidiuretic hormone (ADH). This in turn is carried by the blood to the kidneys and signals to them to reclaim more water from the primary urine. At the same time, you become aware of a sensation that we know as thirst and which impels us from within to drink. According to Tinbergen, this would be the inner or ultimate cause. If we were being looked at by an outside observer at this moment, they would at first notice absolutely nothing. Only when we spot a tap does a proximate cause come into the mix, and it can be observed how we turn on the tap and drink.

When a dog feels unwell because it has eaten too much or something unpleasant, it presumably experiences a 'thirst' for grass. It doesn't begin by suddenly gnawing at a green carpet, but waits until it is taken on its next walk, where it will probably be exposed to the proximate stimulus of grass. We can then observe it eating the grass. Just as we, according to our appetite, opt for food that will really hit the spot for us, so many animals that self-medicate in all likelihood have a comparable mechanism at their disposal. We therefore shouldn't find it at all hard to understand this.

My dog Darwin was born on a farm in Mecklenburg and, until we acquired him, he knew nothing but the few square metres of the farmyard. He came to us as a puppy and so I can rule out the possibility that he could ever have learned from another dog that he needed to eat grass when he had stomach pains. That's what he did, though. Just as we don't have a clue why we feel like eating this or that food at any given moment, so Darwin doesn't know that grass will help him. His behaviour is genetically prescribed or predisposed. A stimulus triggers a cascade of biochemical and neurological processes, and at the end of this stands the need to eat grass. In the event he feels unwell, he theoretically has an unending number of possibilities; he could gnaw at the bark of a tree or eat soil, but at the moment he feels queasy, it's simply easier for him to eat grass.

Perhaps the following example will serve to highlight genetic predisposition more clearly. If, for instance, a child comes from a musical family, he or she will find it much easier learning a musical instrument than a child whose forebears have never played an instrument before. Of course, a musical child has no idea either why it prefers playing music to playing football.

Ants also treat themselves on a comparable cognitive basis. Ants, but also other insects throughout the world, are susceptible to a fungus (*Beauveria bassiana*) that devours them from the inside. This fungus has even been used as a biological form of pest control. Selective pheromones (see 'Pheromone parties') are used to lure certain pests into a trap, where they are infected with the fungus. If they are social insects, the stage is set, as they go on to infect all the rest. That is of course a powerful evolutionary pressure, and so it is perhaps not surprising that they have developed defence mechanisms. A recent study therefore showed that ants affected with the fungus ingested more foodstuffs that contained free radicals. These foodstuffs included the cadavers of dead animals and the sap secreted by aphids, which the ants 'milked'. Here's another great enigma: ants genuinely do live in symbiosis with aphids, which they protect from predators and even carry to fresh plants. In return, the ants receive a sugar-rich sap, which the ants really do 'milk' from them.

The sap contains not just sugar, however, but also a high proportion of free radicals, which the plants originally use to guard against being eaten by pests. Because the aphids suck up the sap exuded by the plant, they naturally also take in these protective substances at the same time. In turn, the ants now use these free radicals as a fungicide to protect themselves against the aforementioned fungus. Watch out, though – no medication comes without its side effects. Free radicals make no distinction between friend and foe, and so the treatment is extremely deleterious to the general health of the affected insects. But even if the ants do not survive their self-healing, they have at least ensured that the fungus dies with them and does not infect their whole colony. As for the disease itself, spreading throughout the ant's entire body, it could well be something out of a science fiction B-movie. Even while the poor animals are still alive, the fungus keeps growing from their body orifices, and when the ants are finally dead a white fungal fluff oozes out of their joints and eye sockets.

Besides, free radicals do not just represent a danger to this fungus; we humans can also be affected by them, and many diseases such as cancer, Parkinson's lymphoma or vascular diseases are attributable to them. What's really perfidious is that we even produce them in our own bodies, as they are in large part degradation products of our metabolism.

But what exactly do these free radicals do? Put in very simple terms, they are chemical bonds that are missing an electron. In order to obtain it, they react with organic bonds that are important to us, and which as a consequence of this reaction are frequently unable to perform their task any longer. They destroy our cell walls, alter metabolic pathways by affecting the enzymes involved, and even damage our genetic material. Fortunately, there are antioxidants that, to put it in graphic terms, throw themselves in front of the radicals and react with them. Their molecular self-sacrifice is our salvation, and thanks to their abundant concentration in fruits and nuts, we are well equipped by Mother Nature to combat them.

But to come back to our ants, in one experiment they were presented with different sources of food: one option involved normal food, while the other had a higher concentration of free radicals. You'll have guessed the outcome already, I feel sure: only diseased ants fell upon the food with the free radicals, whereas healthy insects gave this intrinsically unhealthy food a wide berth.[177]

A while ago I tried fasting. One of the support measures involved me eating healing clay. Naturally, this intrigued me: why should I eat earth when I was fasting? So I did some research and discovered that eating earth is a widespread and long-established practice and that animals also indulge in it. It even has a name – geophagy. It sounded a bit horror-movie to me, I must say, but it suddenly took me back to my childhood and I could remember my grandma telling me: 'Don't worry, a little bit of dirt won't kill you.' I recalled a visit to my aunt's farmyard, where she explained to me that her hens ate dirt because the little stones in it helped them to grind up the food in their stomachs. And it's true; larger birds eat bigger stones, while smaller birds prefer a fine gravel.

My healing clay, however, was in the form of a very fine powder and had nothing to do with stones and digestion. I searched around and came across a fascinating publication from 1999, recounting an experiment that had been conducted with parrots.[178] It's no secret to bird lovers that parrots and other birds occasionally eat soil, which prompted the researchers to pose the same question I'd been asking myself: why? The answer is inspired, and also explains why healing earth is so beneficial during fasting. The researchers ascertained that the parrots in the test ate plants that were poisonous to them relatively frequently. In the final analysis, this came down to an old trick of nature.

After animals left the water and began to become land-based, they developed a great liking for the positively inexhaustible supply of plant matter that was available to them there. As a consequence, the plants evolved countermeasures, growing thorns or developing toxicity so that they would be unpalatable. The tobacco plant, for example, is so poisonous that it is only attacked by the caterpillar of the tobacco hawk moth (*Manduca sexta*). In turn, the caterpillar itself has hardly any predators, since its body contains the neurotoxin nicotine in high concentration, and even a small caterpillar could be fatal to a person. Yet parrots regularly included in their diet plants that were actually toxic to them.

The researchers realized that ingesting very fine soil had the effect of detoxifying the bolus of digested food and bile in the parrots' gastrointestinal tract. The microscopically small and extremely porous lumps of earth effectively absorb the toxins. So the idea of eating healing clay during a fast wasn't so silly after all, since the large surface area of the powder-fine particles was capable of attracting and retaining many substances, including toxic ones. Ultimately, the parrots were practising a form of preventative medicine that stopped them from being poisoned by the plants they ate.

Many birds, and also mammals, use other animals to prevent themselves from being infested by parasites in the first place or to actively combat such an infestation. 'Anting' is a widespread behaviour among birds and involves them settling on an anthill and picking up individual ants in their beaks. Naturally, the ants aren't exactly delighted at this and squirt out methanoic acid, also known as formic acid (from the Latin *formica*, 'ant'). However, the therapeutic effect of this widespread bird behaviour had been disputed, given that formic acid, in the concentration that would be obtained from just a few ants, would hardly have any effect in deterring parasites on the bird.[179] There are also reports of black lemurs (*Eulemur macaco*), which only occur on the island of Madagascar, biting into poisonous millipedes and then rubbing themselves all over with the insects' toxic secretion to protect themselves against parasites.[180] You can watch this behaviour for yourself in a clip from a BBC documentary that has been posted on YouTube (from 1:40 onwards).[181] The baby lemur in the film has clearly taken too enthusiastic a bite, as its behaviour seems to be affected by the toxin. Observations like this have prompted numerous discussions about drug consumption in the animal kingdom – many people found the sight of the small lemur sitting in the fork of a tree and drooling and rolling its eyes too cute and funny.

I could go on to cite many further examples. For instance, there is an anecdote concerning a well-known elephant researcher by the name of Holly T. Dublin, who one day saw a pregnant elephant displaying some remarkable behaviour. The pregnant female left her herd in order to walk 20 kilometres (12½ miles) to feed on a plant that wasn't normally in her diet. Soon after eating her irregular meal, she gave birth to her calf. Apparently, the plant in question is used by native people in the region to help induce labour.[182] In another example, a female spider monkey self-medicated with phytohormones in order to regulate her fertility.[183]

The list of extraordinary examples could go on and on ad infinitum. But what particularly interests me from the point of view of behavioural biology is whether animals like our dogs are following an internal impulse or whether they have learned how to use particular substances. Unfortunately, it is extremely difficult to design experiments to ascertain this, so the question must remain open for the time being. But maybe a report on sparrows collecting cigarette butts and building them into their nests should provide us with some food for thought.[184] It seems this is by no means a birdbrained thing to do, since it helps protect them from parasites. Maybe this doesn't strike you as especially remarkable, until you begin to consider how long there have been cigarette butts around on our streets and how many generations of sparrows it would take for such behaviour to occur in evolutionary terms. Given the impressive cognitive abilities of birds, I am inclined to credit them with having learned to use cigarette butts in this way.

The retired veterinarian Benjamin Hart from the University of California in Davis is considered one of the pioneers of 'animal medicine' in the sense that we have been discussing. In an important article written in 1990, Hart defined criteria that in his view needed to be fulfilled for us to be able to talk about animals practising medicine.[185] One factor was whether animals deliberately ingest active substances. I think we can agree that the examples we have examined so far suffice to prove that this is the case. Another point concerned physical treatment. A monkey delousing another would fall into this category, since it is not just a question of ridding one's partner of parasites but also about ensuring that the parasites can be prevented from transmitting diseases.

Hart's next two criteria were a little more contentious, but seen from the correct perspective I believe these can also be seen to have been met. They concern quarantine and inoculations. In a more recent paper, from 2011, Hart analysed a range of publications on primate research in the wild, some of which were very old.[186] These reported that sick animals were ostracized from their social groups. The extraordinary thing about this was that it happened without any physical contact taking place. Hart was therefore of the opinion that his quarantine criterion had been fulfilled. But what about inoculations?

Anyone who has children has at some stage wrestled with the question of whether to get them inoculated or not. As I've already mentioned, my doctoral thesis was about dolphin therapy. According to their parents, many of the children who took part in this programme had suffered harm through vaccinations. If you find yourself agonizing over the matter and suddenly call to mind these severely handicapped children, it's incredibly hard to come to a rational decision. Nonetheless, I followed statistical evidence and accepted the extremely small risk of a serious disability occurring in return for immunity against a whole range of diseases, some of them deadly.

Not all parents jump that way. So-called chickenpox parties are very popular among this group. You meet up with other parents and their children and visit a family that has gone down with the disease. The party hosts generously donate their germs to their guests, and in an ideal scenario the children come down with a mild strain of chickenpox. In similar fashion, if one is to believe the old reports on primates cited above, immediately after their birth and also later on, baby monkeys were hugged and cuddled by one member of the group after another. Hart saw in this a natural form of immunization, since it exposed the young animal to the various germs carried by the different group members. Accordingly, this criterion was also fulfilled.[187]

Here's a clever thought to end this chapter on: for Hart, strategies to combat diseases are just as important as behavioural strategies to avoid being eaten. This is a crucial point, which emphasizes how great the evolutionary pressure was to develop behavioural strategies against pathogens.

VI

Sentimentality

IN OUR RATIONAL, economically driven world, the term 'sentimentality' has decidedly negative connotations. It describes an
excessive tendency to indulge in feelings that are generally considered negative.

Human language has hundreds of terms that express emotions
– from abhorrence to zealotry. It is possible to apply a number of
these terms to particular manifestations of behaviour. For instance,
aversion can be observed through one animal avoiding another. But
then the question arises as to why this happens. Perhaps one of the
animals hates the other? Yet how to test for hatred?

Cogito, ergo sum – I think, therefore I am – is famously the
principal tenet of the French rationalist philosopher René Des-
cartes. Yet would it also be valid to say 'I feel, therefore I am'? Do
feelings have anything to do with thought?

In some ways they do, for sure: when we have to make really
important decisions, we consult our 'gut feeling'. We ask ourselves
how we feel about the apartment we're looking round with a view to
buying. Does the amazing view it affords and the great feeling that
engenders in us justify the high asking price, or would a basement

flat have the same effect? Should I accept a wonderful job offer at the cost of having a long-distance relationship, or should I stay close to my beloved and forego advancing my career? Which makes me happier? And is happiness even a proper yardstick?

Hundreds, if not thousands of philosophers and psychologists have treated the subject of emotions, yet there is still no generally recognized definition. And I'm far from qualified to make any significant contribution of my own to the debate, although I would like to try to present some useful thoughts here from the perspective of biological evolution. One thing is certain, however: a complex existence would be unthinkable without that facet we call emotion.

27. The interface

For most people, our behaviour is associated with the effect of our nerve cells. That's certainly true in purely biological terms, but even then only partially. We possess a second opposing yet equivalent system of guidance and regulation: namely our hormonal system. It is this system that makes us tired in the evenings, for instance (melatonin), and ensures that our blood doesn't become too sugar-rich and viscous (insulin). Alongside this and countless other physiological control circuits, hormones are also responsible for our social behaviour, as we have already observed through the example of oxytocin.

Unlike the effects of our nervous system, hormones enjoy a certain degree of freedom. Fat-soluble hormones can pass through (cell) walls as if by magic, exert their influence directly on the cell's nucleus and so ensure that particular sequences of our DNA are activated. Other hormones are reliant upon being carried around our body in the bloodstream and require corresponding receptor molecules to be present on the surface of each individual cell. A third group works directly within our nervous system. This is where those brilliant structures called synapses come into play. As outlined above, the signals in the nerve cells are transmitted electrically. Hormones have no access to this process. But when a signal is transferred from one cell to the next, then the hormone's hour is at hand; yet this is also the interface where drugs have their effect. For in the synapse gap between the two nerve cells, they assume the function of neurotransmitters – indeed, they become neurotransmitters in their own right. It is at this point where they intervene directly in our thought processes, and depending on the type of synapse their intervention can be very specific. The wonderful thing about this system is its high degree of autonomy. We don't have to think about it; on the contrary, hormones generate thoughts for us. Through their effect within the synapses, they involve themselves in our thoughts and help us to decide what feels right.

I could now rattle off an exhaustive list of the effects of various hormones, but all that would achieve would be to scratch pointlessly at the surface of the subject. So instead I'd like to examine in greater detail a hormone/neurotransmitter that we've already had cause to mention several times: dopamine.

28. Dopamine – flooring the happiness pedal?

Perhaps you'll recall my mentioning the so-called reward system with regard to play behaviour and birdsong. To be precise, what I was talking about there was the mesocorticolimbic dopaminergic reward system, a network of neurons that branches out to encompass many areas of the brain. This system holds out the promise of yearning, desire and blissful fulfilment.

In the 1950s it was a common practice among behavioural biologists, psychologists and neurologists to insert electrodes into the brains of animals and observe what happened when electrical impulses were passed through the adjacent nerves. When, in 1954, the American learning specialists James Olds and Peter Milner from McGill University in Montreal inserted electrodes in a particular area of the brain – apparently by mistake – they made a remarkable discovery: the rat that was their test animal was enraptured by every electrical stimulation. Intrigued, they proceeded to attach electrodes to this 'wrong' brain area of several rats. The rats were then put in a cage where they could operate a lever controlling the electricity supply and administer shocks to themselves. After just a few minutes the animals had worked out the mystery of the lever and began to push it every five seconds. In fact, pressing it became such a pleasurable experience that they forgot all about eating and drinking. Eventually, the rats collapsed in a state of dehydration and exhaustion. The concept of 'reward centres' had been discovered.[1]

But what do electrical impulses have to do with dopamine? One functions electrically and the other chemically. Yet the truly brilliant thing is the interrelationship of both systems at the synapses. For instance, if the osmoreceptors mentioned in the section 'Under the influence' (pp. 252–8) signal a lack of water, and I become thirsty, then among other things a very ancient area of the brain located within the mesencephalon (midbrain) is activated. The area in question is known as the ventral tegmentum. This region consists of neurons that at the extremities of their nerve pathways – in other words, at the synapses of their axons – secrete dopamine as a neurotransmitter. These extremities, however, are located in quite different areas of the brain and that is where their effect is felt. In this way, areas of the brain such as the nucleus accumbens, the frontal cortex and the hippocampus are flooded with dopamine and hence stimulated.

For almost forty years it was thought that dopamine was directly responsible for triggering familiar happy feelings of gratification. But then a researcher came up with the idea of severing the nerve pathways between the ventral tegmentum and the receptor sites. All of a sudden, the animals lacked any appetite. However, they ate with gusto if the food was put into their mouths. All they were lacking was the motivation to search for anything edible or even to make their way to the food bowl.[2]

Actual reward therefore only indirectly has anything to do with dopamine, since ultimately all dopamine does is stimulate the brain areas that produce the body's own opiates and oxytocin. And it is these that at the end of the day ensure pleasant sensations. If the brain has grasped this connection, something astonishing happens. It learns to equate the effort involved in attaining a goal with the attainment itself. Thus, if you're a workaholic, you no longer work to ends like career success or more money – instead, it's enough for you to work just so that you can feel good.

Of course, this also applies to other instances: a bird sings even at moments when it's not trying to find a partner, and likewise adults take great delight in playing even though they have learned all they need to already. And so it is not goals such as reproduction or social status that lead to fulfilment; rather, it is actions that lead to the goal. Once again, here is a mechanism that could not possibly be more beautiful or ingenious. This connection was recognized some 2,500 years ago by Confucius in his dictum 'The journey is the reward'.

Following the recognition of these connections, scientists identified three distinct, fundamental and separate aspects (wanting, liking, learning) of the reward system:

- The motivation system (wanting) with dopamine as transmitter. In this, desire is generated, which guarantees motivation, whereas any impairment of the system ensures inactivity and depression.

- The enjoyment system (liking). In this, pleasure is generated. This is achieved through endorphins and oxytocin.
- The learning system (learning). Here, the connection between desire and gratification (reward) is cemented, as we learn that we are rewarded.

What's more, these processes take place unconsciously, and the wide range of addictions to which we humans are prone clearly demonstrates what limited power we have over our own reward system. This is due to the fact that drugs work directly within the synapses of the brain areas that regulate the reward system (heroin and nicotine target the ventral tegmentum and cocaine the nucleus accumbens). Young people in puberty are particularly acutely affected. Their genetic architecture makes provisions for a higher number of dopamine receptors. As a result of this, they are more adventurous and open to new and more extreme challenges. From a biological perspective, this makes perfect sense, since this same impulsion is also what prompts them to leave the security of the family and experience things for themselves.

An analogy with animals need not be drawn in this instance, given that all our knowledge about these connections derives from animal experiments anyway. Yet we can safely compare our own feelings with those of animals. If we are thirsty or eat something tasty, then we have corresponding sensations. If we enjoy singing, then we know how a song thrush must feel. And when we sit for hours playing computer games, then we can easily appreciate why wolf cubs play games of rough-and-tumble with one another until they're fit to drop.

If the motivation supplied by a reward system is missing, then the result is depression.[3] On the basis of the aforementioned animal experiments regarding dopamine and other messenger substances like serotonin and noradrenaline, pharmacologists have devised pharmaceuticals that can influence our moods. So-called dopamine reuptake inhibitors, which ensure that more dopamine remains in the gaps between the synapses, were initially very effective as antidepressants, though they have not stood the test of time. Even after a short period, they led to strong dependency, and ultimately patients who took them felt like our rats with the electrical impulses coursing through their brains. By contrast, serotonin and nor-adrenaline reuptake inhibitors have been found to work very well, with a far lower potential for addiction, since they do not act directly on the reward system, which one might well also call the addiction centre.

There is one consolation, however, at least for women. From experiments with rats we know that breastfeeding babies acts as a far stronger stimulus to the reward system than that engendered by, say, cocaine.[4] Even so, the effect of cocaine is still 1,000 times stronger than that of alcohol, with around 7,000 deaths annually attributed to the latter addiction in the UK alone.[5]

29. Cold as a fish

Fish are generally regarded as lacking feeling; indeed, they are supposed not to feel any pain even when they bite into an angler's hook. For instance, brain specialist Professor Brian Key of the University of Queensland has substantiated this assertion by pointing to the fact that fish have no cortex, and hence can feel no pain.[6] If we follow this logic, then from a purely formal point of view birds should also not feel any pain, since they do not possess a cortex either, as we saw in the previous section.

I could at this point go into quite some detail on current research into pain in fish, but instead I'll refer you to an excellent recent overview of the subject in the online *Smithsonian Magazine*.[7] For me, the idea that fish cannot feel pain is positively absurd and, with the exception of Brian Key, nowadays hardly any independent researcher can be found who still doubts the ability of fish to experience pain. The overwhelming weight of scientific literature leaves little room for discussion on this matter. I should therefore like to go one step further here. In the chapter 'The thinking apparatus' we have already learned that fish are increasingly being used by those researching processes within the brain. Thus, we use fish brains to understand our own brains better and to develop drugs that will be effective in humans. Here's a fascinating thought: I'm sure you'll agree with me that the human psyche is an extremely complex thing that we still do not fully understand. Undoubtedly, psychotropic drugs have a considerable influence on this entity. They can, for instance, turn a depressed, listless and introverted individual into a friendly, energetic and active person. It's fantastic that we can get to grips with even severe mental disorders through just a few molecules. One in twenty people in Germany and one in ten in Iceland dare to undergo this minor intervention into the biochemistry of their brains.[8]

Whether this is a good or a bad thing is irrelevant, because what I'm really concerned with here is the question of whether you'd imagine that psychotropic drugs would also have an effect on fish. If so, then do fish have a psyche or a mind? Come what may, we have already seen that fish possess a derivative of the love hormone oxytocin, which among us and them alike helps promote stable partnerships.

Remember the experiment with the mirror and recognizing yourself? I mentioned there that many animals react to a mirror by displaying forms of social behaviour. Mostly this is in the form of aggression, since they regard the mirror image as a rival. The most famous example is the Siamese fighting fish (*Betta splendens*), which will battle to exhaustion against its own image in the mirror or a real rival. But if treated with the antipsychotic drug chlorpromazine (thorazine), which was developed back in 1950, even it can get a grip on its aggression and turn into a peaceful tankmate.[9] What's more, given that this experiment was conducted over thirty years ago, it's a mystery to me how people can still today raise any serious doubts over fish having the capacity for feeling.

Yet you don't even have to take recourse to psychotropic drugs to calm a fish down. In nature, stroking can also help. Some fish that have been stroked show a reduced level of cortisol and hence a lower level of stress.[10] Cleaner wrasses (*Labroides* sp.) make use of this phenomenon, since they are able to carry out their duties more efficiently when their large clients are not thrashing about nervously. This is not an instance of cooperation but of symbiosis; those involved gain mutual advantage from the relationship. Symbiotic relationships exist even between simple organisms, such as fungi and algae, which together form the green lichens on rocks. However, symbioses should not be confused with instances of cooperation. In the former, two organisms of different types have different objectives and exploit one another. In lichens, for instance, the fungus provides stability and retains moisture, while the algae generate glucose from sunlight and use this to feed the fungus. Both partners have learned to specialize over the course of evolution and this specialization is stored in their genes. In a cooperative relationship, on the other hand, two individuals of the same species work towards the same goal. The interaction is not genetically programmed, but is a cognitively directed behaviour. Cooperations are not general patterns of behaviour but rather are an exception, since they presume a certain degree of comprehension of the situation and a process of alignment, as well as rules of behaviour that must be observed. As we have seen, a herd that gathers its young into the centre to protect them from predators does not constitute an instance of cooperation. This behaviour comes through selection. An example of true cooperation would be keeping lookout among ground squirrels.

Is it possible to imagine such cooperation among fish, with one keeping lookout while the other feeds? Could there really be such a thing as cooperative behaviour among fish, based on the principle of mutual benefit? It's hard to credit, but recently it has been found that rabbitfish (*Siganus* sp.), colourful inhabitants of tropical reefs, engage in cooperation. This has an enormous advantage, as rabbitfish that lead a solitary existence can quickly become a tasty meal if they swim too deep into the coral outcrop and fail to notice predators lurking there. This risk is minimized through cooperation. One fish keeps watch while the other swims deep into a rocky crevice in order to exploit the abundant but underused sources of nutrition that are found there.[11] In behavioural biology, this form of behaviour is referred to as vigilance, and was once thought to be confined to birds and mammals. We are yet to discover whether this behaviour among fish is, like their care of their young, also influenced by oxytocin. I wouldn't be surprised, as it is undoubtedly a social behaviour based on trust. It's also unclear whether fish have alarm calls like the ones used by ground squirrels, though even that wouldn't surprise me.

When one also considers that the question of partner choice among humans is being investigated through the study of sticklebacks (see the chapter 'Pheromone parties'; see pp. 32–6), one would surely have to concede that fish are far closer to us than we have previously assumed.

30. Rats – nature's party animals

You'd scarcely credit it, but rats can laugh. They laugh when they are tickled and during play. In addition, laughing rats show a preference for other laughing rats, and tend to shun sourpusses.[12] However, the noises they make lie beyond the frequency range that we humans can hear. Like mice, dolphins and bats, they use ultrasound, so maybe that's why we've never heard them laughing at us. Dogs also laugh, incidentally, and if the sound of 'laughter' barking is played over loudspeakers at dog rescue centres, it helps relieve stress.[13]

So how come I didn't tell you all about this in the chapter 'The hedonistic society' (pp. 105–19)? Quite simply, the subject under discussion here is feelings, but not where they come from or what conditions might lead, say, to happiness. Here, we're looking at the biochemical combinations that make us feel happy. Consequently, it's wholly irrelevant whether we have cause to be happy. On the contrary, it suffices to know which biochemical molecules engender happiness in us. Precisely for this reason, the laughing rats are of immense significance to the pharmaceutical industry. The underlying biochemical processes are so similar among rats and humans that researchers in this field are optimistic about developing drugs to combat depression. To this end, the happiest rats, namely those that in comparison with others spend the most time laughing, are being selectively bred.[14] Experiments are then conducted on them in order to find what substances cause the rats to stop laughing; alternatively, the researchers have tried severing those nerve fibres that they think might be significant. Ultimately, once the cause and effect of the rodents' laughter are known, then scientists can set about manufacturing drugs that reinitiate these processes.

The possibilities, though, do not stop there. The greatest potential lies in combatting violence. We have known since the 1950s that violence has a great deal to do with the loss of empathy and sympathy. If I can no longer intuit the thoughts of my victim or imagine what he or she is feeling, and so cannot envisage the pain they will go through, then there is nothing to restrain me from committing my act of violence. I am simply incapable of imagining the deleterious effect my actions will have on others. This connection was discovered soon after the Second World War, when it became evident that the many parentless children who had grown up in care homes clearly felt nothing for their victims in cases of bullying, and that instances of violence were growing among such children. This gave rise to attachment theory[15] and the concept of basic trust.

Around four years ago, the theme of basic trust became the factor that loomed largest in my life. Every time our twin infant sons began crying, it signalled the start of a crescendo of noise that tested their parents' endurance to the limit. For their part, our own parents just shook their heads and offered helpful comments like: Oh, don't exaggerate, crying's just strengthening their lungs, and what's more the little ones need to learn how to cope on their own. According to the theory of basic trust, constant attention from an attachment figure or figures will result in emotional security. From this, children will subsequently develop a confident attitude towards other people and their environs. If this attention is lacking or if trust is undermined, what will result is a fundamental mistrust and the child's dealings with others and their surroundings will often be characterized by anxiety. The child only trusts itself and struggles to make headway. This lack of affection robs a person of their capacity for sympathy and empathy and violence grows as a result. Those who are affected inhabit an Old Testament world ruled by the principle of 'an eye for an eye', and fail to heed Mahatma Gandhi's rejoinder '…makes the whole world blind'.

Although broadly speaking crimes of violence have been on the decline latterly in the Western world, violence remains one of the most pressing problems in human societies worldwide, and so understanding the biochemical root causes is an attractive and also lucrative field of research.

But what is the difference between proportional violence and a dangerous escalation? Undoubtedly, most people are capable of a modicum of violence, and from personal experience I know how quickly I can metamorphose from a peaceful pedestrian into a raging monster. Adrenaline makes this possible. But my violence has its limits, and I would never go beyond a proportionate level of force, and even then only resort to violence in self-defence. What stops me from going further is not only my sense of justice, but also a feeling of compassion that operates according to the precept: who knows what happened beforehand to my adversary or what has made him the way he is? Little wonder, then, that research has identified sympathy as being the principal means of countering escalating violence.

Now, you can of course only study sympathy in animals if the animals are capable of compassion in the first place. Behavioural biology talks in terms of prosocial behaviour. Do you recall our experiment with the grapes, in which the chimpanzee that was favoured also did without its grape when it noticed that its companion had not been given one? Or the rat that freed its friend from the trap? Both gained nothing from their behaviour. On the contrary, they both lost out as a result – in the chimp's case on grapes and in the rat's on time it could have spent doing something else. They acted selflessly out of compassion. Because experiments involving great apes are now ethically controversial and also because rats are perfectly capable of displaying compassion in tests, current research is focused on the latter.[16] Researchers are especially interested in the amygdala, an almond-shaped set of neurons located deep in the brain's medial temporal lobe, which play a key role in social interactions.

Looking at some of the animal experiments that are still being conducted today, especially on rodents like our laboratory mice and rats, the first question that springs to my mind isn't: do these animals have compassion? Instead, I find myself asking whether the experimenters have any sympathy. In scientific terms, however, even this behaviour can be plausibly explained. The more advanced a creature's cognitive development, the greater the capacity for cognitive processes to also influence its actions. Of course, this holds good not only for human behaviour in many animal experiments and in factory farming, but also our conduct towards one another. Otherwise, how could one possibly explain many war crimes and organized crime?

VII

The pinnacle of creation

A T THE BEGINNING of this book I promised you it wasn't my intention to topple humanity from its throne, and nor indeed have I done so, despite the fact that we have seen numerous examples of the ways in which we humans do not differ fundamentally from the rest of the animal kingdom in either our thoughts or our emotions. Nonetheless, there is a very rational explanation for the success and dominance of *Homo sapiens*.

31. The human USP

We humans are a truly remarkable species. Perhaps you'll recall the insight that we are the only species who allow ourselves to be tormented, exploited, debased, humiliated and insulted. That was the gist of the chapter 'BDSM' (see pp. 36–8). This behaviour could only be explained by the fact that a few human individuals are capable of defying circumstances that would normally evoke an immediate flight response. Leaving aside any value judgement, this is truly an impressive mental achievement, since such behaviour is only possible if I am cognitively in a position to transcend internal conditions that control my behaviour in the general run of things. We humans can consciously decide for ourselves to drink nothing for an entire day or to eat nothing for four weeks, although we would normally sit down at table to eat with our families three times a day. We humans also have the capacity to disregard dependencies. Even if we are suffering under the influence of a mis-directed reward system (see 'Sentimentality'), we can overcome our alcohol or nicotine dependency any time we choose. Granted, this may require therapeutic support, and many addicts lapse back into their dependency, but it is theoretically possible all the same.

Yet we also have our limits. Although I can hold my breath for a reasonably long time, I am not capable of holding it for so long that I suffocate. Certainly, as a basic premise we are perfectly capable of killing ourselves, and in that moment are even able to ignore our deep-seated instinct for self-preservation, yet we are still incapable of taking our own lives simply by holding our breath and asphyxiating ourselves. Dolphins, on the other hand, might well be capable of that feat. For instance, former dolphin trainer Richard O'Barry reported that one of the animals used in the *Flipper* television series suffocated itself in his arms. Even though such an anecdote cannot be corroborated, many such stories exist. Despite the fact that suicide among animals is regarded by the scientific community as unproven speculation, twenty leading brain researchers recently called for the neurological mechanisms of suicide to be investigated.[1]

They have proposed that animal experiments be conducted in order to gain a better understanding of the phenomenon of 'suicide'. They are not interested here in whether animals are capable of committing suicide, but rather in throwing more light on the phenomenon of human suicide, which nonetheless only comes fifteenth on the list of the most common causes of human mortality. Particularly prone are individuals who combine within themselves a number of valuable cognitive achievements. Thus, people who take their own lives have a propensity for self-reflection far exceeding the average and also possess abstract and complex emotions. A number of different structures in the brain (such as VMPFC, DMPFC, pgACC, sgACC, together with the amygdala and the insular cortex) are involved in these capabilities, and it is the interaction between these different areas of the brain that scientists wish to investigate further using the model of the animal brain.[2]

In any event, where suicide is concerned, we are dealing with a form of behaviour in which a high degree of cognitive development enables an individual to defy other internal control mechanisms.

However, we humans display one further strange characteristic. No other animal integrates itself so willingly into the mass. In other words: only we humans ignore our own needs in order that we might go on cooperating. This behaviour begins in the kindergarten and ends in high diplomatic circles – for instance, when Chancellor Angela Merkel holds a summit with President Vladimir Putin of Russia, even though she has absolutely no desire to do so.[3] The scientific term associated with this behaviour is 'normative conformity'. What is meant by this is subordination and renunciation in favour of belonging to a group, with its attendant rules. Even though it would be interesting, unfortunately neither Angela Merkel nor Vladimir Putin are about to make themselves available for experiments in behavioural biology, and so the scientists are obliged to use other individuals such as chimpanzees, orangutans or children of kindergarten age. The question is: would an orangutan also hold talks with Mr Putin if these were called for in the context of a shared societal convention? And the answer is: no. Unlike kindergarten children, who put their own interests aside, orangutans and chimps would not talk to anyone they didn't like. But let's take a closer look at the experiment.

Imagine you're sitting in front of three boxes with holes in them. After briefly checking them out, you've discovered that a small reward tumbles out of one of the boxes when you throw a ball into the hole. You're the only kid in your kindergarten group to have sussed this trick. Now three of your friends appear and are also allowed to try their luck. Of course, you know the trick, and ideally you'd like to help them, but instead you've got to stand back and wait until it's your turn again. What would you do if all of your friends ended up pitching their balls into the wrong hole?

Chimpanzees and orangutans would simply step forward again, and in happy anticipation of the reward, cast their ball into the correct hole once more. Human children, on the other hand, only do that if they don't think they're being watched by their friends. If they feel themselves to be part of a group, they will deliberately make the wrong choice and consequently get no reward.[4] The advantage of such behaviour is that you won't be left alone with your reward outside the group. Your behaviour won't provoke any feelings of covetousness or envy and you'll remain part of the small community because you won't have set yourself apart by virtue of the reward you received.

When Mrs Merkel talks to Mr Putin, she's behaving in a very similar way. Although it runs counter to German and European interests to lend diplomatic succour to an autocrat and warmonger by agreeing to a visit, the Federal Chancellor's behaviour ensures that dialogue is maintained and that all those involved get the impression that they are part of a common network.

What nonsense, you'll be thinking, to present compliance and conformism as great boons to mankind. Isn't the exact opposite the case? Don't precisely those people who take a greater risk and distinguish themselves by doing something extraordinary enjoy more success? Of course, you're correct in this observation, for as we saw in the section on 'Personality' (pp. 204–17) there will always be individuals who break new ground and benefit from doing so. Ultimately, this behaviour benefits the entire community. And it doesn't matter one iota how many maverick individuals fall by the wayside in the process; the key thing is that one is successful and that others can safely follow in his or her footsteps.

Thus, the value of individuality is not disproved by the experiment in human conformity outlined above. However, if our society consisted solely of headstrong individualists, it would soon disintegrate, as everyone would be pulling in different directions. But the precise opposite is in fact the case. The great majority of people behave in an extremely group-conformist way. This begins with football fans, but also pertains in our professional lives and could be regarded as manifesting itself in its most extreme form in groups like the Taliban. Within a community we are strong, and we are prepared to forego our little egotistical needs to ensure that.

It is not wholly improbable to see this tendency to self-sacrifice and conformity as having paved the way for our high degree of willingness to cooperate. The renowned Leipzig behavioural scientist Michael Tomasello, for instance, has spoken of human beings as the animal that says 'we'. Of course, many other animals also cooperate, but none on the same level as humans or with the same intensity. If we're setting out to try to discover a human 'unique selling point', we have certainly hit on something here. No matter whether it's to do with assembling rockets that take us to the Moon or inventing a mobile phone operating system called Android, we do it in cooperation with others. As we saw from the example of CAPTCHA (see 'Patent office or open-source?'; pp. 58–73), we can even cooperate with millions of other individuals and through this cooperation tap into the knowledge of individuals who are long since deceased. Even our language came about through cooperation and is, so to speak, a by-product of our social mode of living. No person, however clever, would be capable of developing a language on their own, though he or she would still be theoretically able to devise the theory of relativity.

In my view, this pronounced degree of cooperation among humans is the sole, truly major difference between ourselves and other animals. As humans, we live in a world that has come into being over generations. Our cities, our infrastructure, scientific endeavour and culture are unlike anything else we know. It is nothing short of fantastic what 'we' have built up. And yet as individuals we are not even capable of mending a broken hair-dryer. To be precise, as individuals we are probably not even able to ensure our own survival, for how many of us nowadays even know how to forage food for ourselves?

But if we look at the aptitudes and achievements of a single individual, it is hard to discern any difference from other cognitively highly developed animals. Many species can think abstractly and logically, and they can even reflect on their knowledge and thought processes. They possess a biography and self-awareness and plan their future on the basis of individual characteristics and experiences. In addition, there is much evidence to suggest that many vertebrates have a very similar emotional life. It almost seems as though it would be more apposite to identify a degree of humanization at work here, rather than positing a strict dividing line between us and other animals. Given all these similarities, isn't it much more probable that we all 'tick' in pretty much the same way? And wherein might this great difference be supposed to reside? The Israeli historian Yuval Noah Harari has a clear response to this, and I strongly recommend you take a look at his TED Talk.[5]

Regardless of the numerous experiments and findings we have thus far cited, I would like to elucidate this standpoint from another couple of perspectives. On the one hand, I propose we turn our attention to the evolution of the brain over the course of time and on the other to neurological research into mental illnesses.

Brain development

Truth be told, our nervous system is an example of faulty design. If we were to compare a single nerve cell (neuron) with a hydroelectric power station, then we would have to say that the retaining dam wall is riddled with countless holes, through which huge amounts of water flow constantly without being put to good use in driving the turbines. This is exactly how a nerve cell functions, only it is not water flowing here but rather sodium and potassium ions. These too are pumped against the flow direction with a great expenditure of energy. The charge thus generated enables, in the case of stimulation of the nerve cells, the stimuli to be transmitted as signals. However, the neurons are not fully impermeable to the sodium ions, meaning that there is a steady 'inward leak' of positively charged sodium (Na^+) ions – in other words, a kind of short circuit. Every individual nerve cell, no matter whether it is in use at the time or not, therefore constantly consumes energy, and a living organism must be able to sustain this. We humans, who possess in the region of 86 billion nerve cells in our brains,[6] are carrying around with us an organ that for most of the time is only acting as a heater. A multicellular organism cannot do without stimulus transmission, but why did brains, with their multitude of nerve cells, come into being? This kind of control centre naturally has its advantages. It enables me, for instance, to compare sensory impressions, store important information and possibly even carry out simple mental tasks.

But one doesn't need a large and intricately constructed brain to perform such tasks. In my opinion, the most plausible explanation is the so-called social brain hypothesis. This harks back to an idea that is some forty years old.[7] According to this hypothesis, large brains did not develop in order to facilitate particularly intricate thought processes, but in order to manage our complex social lives. The formation of groups and the cooperation this enabled have manifold advantages for the individuals within them, but this form of coexistence also presented new challenges.

For example, how can we ensure that individuals don't cheat their way through at the expense of the group? That is of course only possible if one clocks an individual as a cheat, spots their behaviour, and behaves accordingly towards them. Yet, as we have had cause to note throughout this book, these supposedly simple achievements are based on a variety of cognitive feats that are anything but trivial. At the end of the day, the computing power necessary could only be managed by large, intricately constructed brains.[8]

But for the sake of completeness here I should mention that the social brain hypothesis also has its critics. For instance, solitary bears scored well in problem-solving tests,[9] and in tests of self-control there was found to be a stronger correlation between brain size and diet than between brain size and social life.[10] In my view, though, these results do not refute the theory of the social brain. On the one hand, we do not know how the animals that were studied evolved; it may be that their ancestors were extremely social. On the other hand, only a limited number of test procedures were employed and it is logical that some species will perform better than others in a particular test. But who knows, perhaps hunting for food and social living played an equal role in the evolution of large brains?

Whatever the case, if I as a human being consider whether I can stand a particular person or if someone has treated me unfairly, then I am presumably thinking the same thoughts as a rat would. It too reflects on itself by means of metacognition, puts itself as an empathetic creature in another's mind, and has a conception of fair and unfair behaviour. If one bears in mind the observation on dialects among rodents, it is perfectly possible that rats might also be capable of distinguishing individuals. This all sounds like a pretty bold assertion, I grant you, but I would certainly endorse this kind of analogy where dolphins are concerned. However, as we shall presently see, I had good reason for choosing an analogy with a rodent.

Mental illnesses and psychotropic drugs

In the chapter 'Rats – nature's party animals' (pp. 278–82), we noted how important studies on rats have been for research into depression and violence. In a study that was published in 2010, a research team investigated differences in the genetic layout of the nerve cells in the mouse and the human brain. They came to the conclusion that the genes involved scarcely differed at all, and so proposed using mice as research subjects in order to study neurological disorders and the mental illnesses resulting from them.[11] Maybe you're thinking that animal experimentation with mice is nothing new. That's true, except the focus of research has shifted. When animal experiments were carried out in the past, we wanted to find out whether a particular drug would be effective in treating a heart condition, cirrhosis of the liver or a rash, say. It was all about our physical being, not our mental one. Even just a few years ago, the idea of using animals to research mental illnesses was still completely inconceivable. Indeed, why would one have proposed such a thing, given that only humans were deemed to possess a mind or a psyche?

Nowadays, no serious scientist would dispute that animals have a psyche and are capable of experiencing psychological disorders. Just like us, animals can be happy or depressed, and in all probability the breadth of our emotional spectrum is something that developed over the course of evolution. It's perfectly possible that we humans have added certain feelings to this exhaustive list. Thus, I can well imagine that the phenomenon of *fremdschämen* – that is, the 'cringe factor' of feeling embarrassed at others' behaviour – is a purely human emotion. Yet this feeling is extraordinarily complex, and we Germans can pride ourselves on having coined a term for it.

Basically speaking, one can certainly proceed from the assumption that a mouse's psyche is no different to that of a person; otherwise, the drugs that have been developed through tests on mice would not work for us. As we have seen above, psychotropic drugs can also reduce aggression in fish. But what are we to do with this finding?

Let us look for a moment at the history of one of the most important scientific journals in behavioural biology. *Ethology* was founded in 1937 and published the works of such eminent figures in the field as Konrad Lorenz and Nikolaas Tinbergen. It's significant that the journal was called '*Journal of Animal Psychology*' when it was first established. Yes, you read that correctly – psychology. Back then, the behaviour of animals was explained in terms of their psyche. By 1985, academics in the field had come to regard this title as insufficiently scientific, and so the journal's name was changed to *Ethology: International Journal of Behavioural Biology*. The inner mental state of animals was disregarded because it could not be researched. Instead, the focus now turned to observations of behaviour, which in purely formal terms had nothing to do with the mind in the black box of the brain. In each case, the most simple logical explanation was sought for a particular behaviour, and animals were seen as little more than biological robots.

The way human societies treat animals is based in large part on this school of thought. Animal welfare legislation was drafted by experts who proudly identified themselves as behaviourists, by which they meant that they were rationally thinking scientists who were unmoved by any inappropriate emotions or ideas of humanization. But now that hardly any serious scientist would dispute that animals have a mind or psyche, perhaps it is time that *Ethology* reverted to its original title.

It's my contention that a great deal more than that should happen, however. We need to radically rethink our treatment of animals and bring it in line with the findings that are currently emerging. It's difficult to foresee what this will mean in detail and what the proper moral ramifications might be. But the fact remains that, as it stands, our treatment of animals, be that in the wild or in captivity, is inappropriate and has evolved on the basis of inaccurate assumptions and simplifications. Even so, there may be a glimmer of light on the horizon.

32. Experimental errors

Scientific findings are commonly regarded as a sound basis on which to make a decision. But even schoolchildren learning the basics of science have it dinned into them that they should be prepared to analyse their experiments for any errors. For this reason, let us now turn our attention to the genuinely extremely important topic of experimental errors.

Misinterpretations

On the very first page of this book, I revealed to you that I made a monumental miscalculation in Florida while I was working on the abstract for my doctoral thesis. I had observed that dolphins in the swimming therapy programme sought out the company of swimmers. As my own data was later to prove, this was quite wrong. To be precise, the opposite was in fact the case; the animals were trying to avoid the swimmers within the confines of the pool. Yet to an observer it looked as though the dolphins were constantly in close proximity to the people. The reason for this misinterpretation lies in the form of human attention. We tend always to direct our attention towards a place where something is moving. After half an hour's observation, we might easily get the impression that the dolphins spent all their time frolicking around the people. But in actual fact the animals simply lacked sufficient space to avoid them. This belated insight, which I only came to after evaluating my research data, came as a real shock to me. Nevertheless, this experience brought it home to me clearly how easy it was to make a mistake. After all, I had already been in Florida for a year studying these interactions day in, day out, but my mistake had been never to question my original assumptions from the pilot study.

Some facts simply elude our attention, and it is only through statistics that we come to a realization of the true situation. Fortunately, I am not alone in having fallen victim to a misinterpretation. Charles Darwin, for instance, thought that chimpanzees were laughing when they pulled back their gums and bared their teeth. In actual fact, this is the way in which a lower-ranking animal signals its fear and submission to a dominant chimpanzee. Many people also think that dolphins can laugh, although they have no capacity whatsoever for mimicry; in truth, their 'smile' is genetically imprinted into their face.

It's really funny when onlookers ascribe emotions to a robot, just because it moves like a dog and wags its tail in a 'friendly' way.[12] The Turing test takes its name from the brilliant British computer pioneer Alan Turing.[13] In this experiment, test subjects communicate with an interlocutor via a screen. When prompted, they reported on the ailments they were suffering from. On the basis of this brief medical history, they were given a diagnosis. Following this chat, they had to decide whether they had been talking to a computer or a human doctor. Most of them were surprised to learn that their doctor was in fact a computer. It's also amusing to watch a video from the 1940s[14] in which test subjects are shown watching geometrical figures chasing one another.[15] Incidentally, there's actually a great deal of money riding on these comical instances of human–computer interaction; the ultimate aim is to design robots that we will look upon as social partners and therefore be more inclined to buy. But these old clips also demonstrate how readily we humanize things and how easily we fall for little tricks.

The thing I'm really driving at here is this: we humans need to watch out that we don't transfer our own yardsticks and expectations. I urge you to keep this in the back of your mind if you follow my suggestion and humanize animal behaviour a little. This is in the truest sense of the word a high-wire act, and every humanization should be conducted on an objective basis. It is therefore in my view wholly appropriate to transfer the human perception of emotions to animals, given that the biochemical basis for this is the same. I also think it is reasonable to draw comparisons when animals produce comparable results in comparative experiments.

False negative results

What on earth is meant by 'false negative results'? Once again, it sounds like a pointless complication, and yet this concept conceals one of the greatest problems facing science. The basic idea is a simple one: if I have a hypothesis then naturally I would like to test it, and I hope to obtain positive results. If the results are negative, though, I must discard my hypothesis. But what if my hypothesis was right, and it was just my experiment that was wrong? If I consequently get negative results, despite my original assumption having been correct? This is where scientists talk of 'false negative results'. Let's examine this using an example we've cited already.

For several decades, the developmental psychologist Michael Tomasello from the Max Planck Institute in Leipzig has been conducting comparative studies of children and other species of anthropoid ape. One of his most important areas of research is the question of a theory of mind in animals (see 'I know that you exist/Theory of mind'; pp. 217–24). He is therefore investigating whether animals are capable of putting themselves into the mind of another. In an attempt to settle this question, he has carried out innumerable experiments over the past few decades. He came to the conclusion that only we humans have the capacity to employ the highest form of the theory of mind, namely the recognition of false belief. In his view, it is only us humans who are able to recognize that other individuals might evaluate an identical situation in a totally different way to us. We spent quite some time examining this question, since ultimately we base many of our moral decisions on our ability to identify, for instance, whether a person did something deliberately – in other words, with intent – or whether they were acting on the basis of impaired reasoning or a miscalculation. For instance, Paragraph 20 of the German Penal Code states: 'Whoever upon commission of the act is incapable of appreciating the wrongfulness of the act or acting in accordance with such appreciation due to a pathological emotional disorder, profound consciousness disorder, mental defect or any other serious emotional abnormality, acts without guilt.' Without the ability to recognize a false belief, it would be impossible to formulate this legal principle or to apply it in practice.

Today, we know that Tomasello was on the wrong track with his assessment and that for decades he had been publishing false negative results. Of course, he did not do this out of deceit, but because his results did not permit any other conclusions. Basically, he was employing the wrong research method. The fact that he was also a co-author of a new publication in which other great ape species were stated to have shown evidence of a theory of mind and an understanding of false belief emphasizes how science ought to operate. It's not about being proved right but about being able at all times to change one's own view of things on the basis of new findings.

Things get problematic when political decisions to the benefit or detriment of animals are made on the basis of the current state of scientific knowledge – for instance, when we make statutory provisions that all vertebrates must be slaughtered painlessly, but exempt fish from this requirement. The person who framed this legislation proceeded from the assumption that fish do not feel pain. On a daily basis, rulings such as this are responsible for causing excruciating pain to millions of animals, and – as a result of neurological processing and the resulting distribution – untold suffering as well.

We encountered a similar situation regarding logical thought. Dogs signally failed this test until it was adapted to suit dogs and consequently the animals passed it with flying colours (see 'Logic'; pp. 175–8). Another example is the mirror test for self-awareness, which to begin with could also only be successfully tackled by animals that were capable of touching their own foreheads.

There are many reasons why an animal might fail to display a particular aptitude in our experiments. Sometimes emotional involvement is lacking, and sometimes a certain degree of aggression; sometimes the reward simply isn't attractive enough, and sometimes the sensation being tested isn't the correct one. A mirror test relies upon visual perception, but what if a pig needs to smell in order to be able to recognize itself, and it takes being marked with another's scent to get a reaction?

Of course, the whole business can work the other way round, since it's also possible for me to obtain false positive results. There is scarcely a scientific paper that does not provoke critics to come to an opposite conclusion. Thus, it was recently claimed that corvids do not after all possess the capacity for abstract thought,[16] and that dolphins have no self-awareness.[17] Inevitably, the critics are then criticized in their turn.

Even so, I imagine that over the coming years we will be treated to a plethora of new and above all surprising findings on animal cognition. We will learn that the way we regard ourselves in our dealings with animals is based in large part on a misjudgement and is ultimately the result of false negative conclusions. The findings on the cognitive abilities of animals presented in this book are relatively new in the main, yet there are many surprising findings that are twenty years old and more. However, you will search in vain to find a record of this knowledge, even in current textbooks for students of biology. Thus, just one of fifty-nine chapters in the 2,000-page standard work on biology, William K. Purves' *Life: The Science of Biology*, deals with animal behaviour. Furthermore, there is not a single example of cognitive abilities of the kind I have cited here in their hundreds. This came as a huge surprise to me too, and so in the spring of 2017, I floated the offer of an interdisciplinary seminar to almost 300 universities, high schools and teacher-training colleges in the German-speaking world. Only a handful of institutions were willing to include such a module on their syllabus in the next year or two. The teacher-training colleges mostly declined, citing the following reason: 'Since curricula are a statutory requirement, it would be legally out of the question to offer a different content in lectures and seminars.' So, is politics responsible for our knowing so little about how animals think and feel?

★

Comparative behavioural research

Comparative behavioural research of the kind conducted by Michael Tomasello, say, is seen as being scientifically extremely precise, and it is hard to escape the compellingly simple logic of this assertion. If I test two different animal species for a particular aptitude using the same method, this enables me to prove conclusively whether the two species share these capabilities or not. If an animal passes a test for logical thinking just as successfully as us, then it can be demonstrated beyond doubt that the animal can think logically. It's all beautifully simple. But as is so often the case, the devil is in the detail, and there is a relatively high chance of obtaining false negative results.

For instance, it may be that the two species being compared receive different food as their reward. Potentially, however, this reward is not equally attractive. Thus, the two-year-old children who took part in Tomasello's conformity test (see 'The human USP'; pp. 283–9) were given chocolate, while the apes received nuts. It also makes a difference if an animal is hungry or not. If this isn't taken into account, we might falsely conclude that hungrier animals are more intelligent. In the current ongoing experiment on the theory of mind among great apes, the situation is emotionally charged and hence made more attractive by the addition of elements of aggressive behaviour. But what if this emotional involvement is not present to the same degree in the species being compared?

Comparative behavioural research also has another serious drawback: it is reliant upon experiments on captive animals. Especially where the investigation of cognitive abilities is concerned, therefore, one quickly comes up against methodological barriers. From the age of two at the latest, human children have a 'common ground' – in other words, a collective self-image when dealing with other people that is based on cooperation. Critics of comparative

behavioural research make the convincing case that human children involved in experiments where other human beings are the
testers have a distinct advantage. They are socialized to interact
with unfamiliar people. Critics also think it highly likely that the
captive animals share absolutely no or only an extremely limited
'common ground' with the researchers and keepers.[18] We should
therefore not be surprised to discover that Tomasello has been
unable to identify any 'common ground' and so takes the view that
there is no possibility of complex cooperative interaction.

His colleague in Leipzig, Christophe Boesch, sees things rather
differently: 'Ignoring most published evidence on wild chimpanzees, Tomasello et al.'s claim that shared goals and intentions are
uniquely human amounts to a faith statement.'[19]

33. Of humans and animals: whale watching versus whale hunting and drive hunting

The relationship between humans and animals is ambivalent in
the extreme. To some people pigs are unclean, while for others
cows are sacred. Some people eat dogs, while such a thing would
be unthinkable for others. There are some people who pay out
thousands in veterinary bills to treat a much-loved pet but who think
nothing of buying cheap pork at £4 a kilo. Everyone has their own
reasons for behaving in this or that way, and most people consider
themselves animal lovers. The whole thing is irrational, and everyone has their own personal demon pricking their conscience, for
as so often brutal exploitation of animals and peaceful coexistence
with them are very close neighbours.

For me, these two worlds collided in a taxi in Cape Town. I was
on the way from the airport to my hotel when something jolted me
out of my daydreaming: 'Where are you from?' came the driver's

voice over the sound of the engine. 'Germany.' – 'So what are you doing here in Cape Town, holidaying?' Yes, good question, what was I doing here in Cape Town? Originally, I'd been scheduled to say a few words at the press launch of a new TV film starring Veronica Ferres, Mario Adorf and Christopher Lambert, he of *Highlander* fame and the screen idol of my youth. My role at the press briefing was to stress that the ecological thriller *The Secret of the Whales* was a work of fiction, albeit one based on real events. The production company had previously asked me to comment on their script and, as far as I was able, to ensure it reflected reality. But in the event, my short slot at the briefing turned into an extended Q&A with a small group of journalists who were interested in the film but also wanted to know more about noise pollution in the world's oceans.

'No,' I told the taxi driver, 'I'm not here on holiday, and truth be told I don't quite know what to expect. But in theory it's whales that have brought me here to South Africa.' What he said next came like a bolt out of the blue: 'Yeah, it was whales that brought me to Cape Town too, thirty-five years ago. I tell you, it was a real boom time until all the whales disappeared. Then they even stopped us from hunting the few that were left. But that's how it goes, I guess; no one looks out for the little guy.' I was wide awake by this stage. Right in front of me I had an eyewitness from the era when they hunted whales! Here was a unique opportunity to chat face to face with a former whaler, and the ride to the hotel would only last a quarter of an hour more at most. In hindsight, dozens of questions occurred to me: How long did it take before a whale died of its injuries? Did you hunt the calves as well? Did anyone have qualms about hunting whales back then? But at the time, all I could think to ask him was what he'd been up to since. 'Oh, not much in all honesty. I did odd jobs here and there, then seven years ago I started driving taxis here in Cape Town, and all the while I could have been cashing in on my expertise. Some of my whaling mates run whale-watching tours these days. They've been raking it in, I can tell you. Now, you'll know that whales aren't stupid. Once we stopped hunting them they came back, and guys who had nothing better to do with their time started taking tourists out on trips... There you go, that'll be 240 rand, please!'

Still in something of a daze, I paid the fare and thanked the cab driver for the interesting chat. What could I have said in response? That the animals hadn't simply swum away, but that whale hunting had almost wiped them out back then? That South Africa was one of the few examples of a place where a whale population had increased at a rate of reproduction that represented the maximum that was biologically possible? What would he have said if I'd told him that whale watching now goes on in sixty countries and that South Africa is number five on the list of favourite destinations? Presumably he'd just have smiled his toothless grin and shaken his head.

South Africa really is a special case, as it's not everywhere on Earth that you can watch whales from dry land. Whale watching is an industry that's still in its infancy, with the first excursions specifically aimed at observing cetaceans beginning in 1955 in southern California. In Europe, this branch of the tourist industry only began in 1983, with trips from Gibraltar. Today, whale tourism has burgeoned into a sector that is worth several hundreds of millions of dollars and has created many jobs worldwide. However, it can only last if it is carried out in a sustainable way. In addition, whale watching can be an incredibly effective way of educating people about environmental issues.

An intriguing example is the still-troubled history of Iceland. By the start of the twentieth century, Norwegian whalers had so severely depleted whale stocks in Icelandic waters that, as early as 1915, the Icelandic government declared a twenty-year moratorium on catching whales. However, hunting did begin again and only ended in 1989. Despite the fact that commercial whaling had been banned at an international level in 1986, between 1986 and 1989 the Icelandic whaling fleet killed an average of 100 fin and sei whales annually on the pretext of 'scientific research'. When the first company on the island of geysers and glaciers started offering whale-watching trips, in 1991, and attracted just 100 paying customers in its first year of operation, no one suspected that that figure would grow over time to over 100,000. Whale watching has therefore become a genuine economic alternative to whale hunting, and whale watchers take a dim view of nearby boats firing harpoons at whales.

The downside of this commercialization of whale watching is of course the increased pressure of competition, and unfortunately it is all too common a sight to see a skipper trying to get right up close to the animals. This all-too-human behaviour needs to be controlled, and above all requires self-control on the part of the whale-watching tour operators. In this context, the tourists also play an important part, as it is only too easy for them to exert pressure on the skippers to take them closer. For this reason, all whale-watching activities should basically come with an educational package. Only when visitors understand why boats should keep their distance from whale pods can a sustainable whale-watching programme be implemented in the long term.

Another negative effect of whale watching, but also of a misdirected educational programme in dolphinariums, has become a real bane for cetaceans, as more and more people feel the need not only to observe whales and dolphins but also to feed and hug them. Ever since *Flipper* was aired, at the very least, we humans have come to regard dolphins as waterborne versions of ourselves, and feel an urge to get as close as we can to them. Our image of them has been shaped by their strangeness, their intelligence, their grace and their apparent cheerfulness, and very few people still see dolphins as a wild animal or as a potentially dangerous predator. The hundred-plus razor-sharp teeth in a dolphin's jaws have already been the cause of some nasty bites, and being bumped by the animal's snout or whacked with its tail has accounted for several broken bones among swimmers.

We now believe that the historical image of dolphins and their depiction in the mythologies of various civilizations are based on a misinterpretation or an over-interpretation of observations of coastal dolphins, which lead solitary lives. There are many instances of sightings of these loners, both historical and contemporary. In truth, however, the behaviour exhibited by these animals is an exception, and could perhaps best be compared with that of human hermits. It is necessarily wrong to extrapolate from this handful of animals and draw conclusions about the behaviour of the entire species. Besides, apart from a very few exceptions, these solitary dolphins can create problems and only a handful of these animals survive proximity to us humans for very long.[20]

Basically, although dolphins are inquisitive, they are also shy and a close encounter in the water, maybe even involving physical contact, is as likely as a lottery win. And so our human conception of dolphins and their portrayal in the media are well wide of the mark.

Finally, I'd like to acquaint you with another exceptional piece of human behaviour. Have you seen the Oscar-winning documentary film *The Cove* (2009)? It follows the former trainer of the famous TV dolphin 'Flipper', Ric O'Barry, on his journey to an idyllic cove in southern Japan. But what happens there is anything but idyllic. Every year there, without fail, the drive-hunting season begins and the cove at Taiji, along with many others in the region, turns red with the blood of dolphins that have been brutally slaughtered. Ric O'Barry's contention is, if we can't call a halt to this abuse, how can we hope to save the entire planet?

The film provoked a huge outcry around the world, and many people came to realize that this bloody slaughter could be ended were it not for the demand by marine parks and dolphinariums all around the world for healthy young female animals – the main market that the fishermen at Taiji were supplying. But Japan is a long way away, so what is the situation in Europe? Surely that kind of thing can't happen here – or can it? The surprising answer is yes, it can. Something similar is also an annual practice in the kingdom of Denmark. On the Faroe Islands, a self-governing Danish archipelago in the North Atlantic, drive hunts of pilot whales and other species are a regular occurrence. The Faroese don't set much store by European Union regulations on animal welfare, and nor do they have to, as their self governing status means that they are not part of the EU.

But as always, the situation is not that simple, and confrontation has in the past yielded little in the way of results. But one thing at a time: the Faroese hunt pilot whales – strictly speaking, these animals aren't whales but instead form part of the dolphin family, alongside orcas and bottlenose dolphins. To couch it in very unscientific terms, pilot whales are a mix between whales and dolphins. They lead an extremely complex social life, like many dolphins, but are larger and display a correspondingly relaxed demeanour. With compact bodies that are well designed to cope with the roughest sea and weather conditions, pilot whales live in groups with a social structure akin to that of orcas. In the case of pilot whales, though, the group is somewhat larger, numbering from twenty to ninety animals. The groups consist of males, females, juveniles and calves, with the older females performing the most important social role, as the group leaders. It is not uncommon to see groups comprising several hundred individuals. They can often also be seen swimming alongside other whale and dolphin species.[21]

Not much is known about their cognitive abilities, as their habitat in the extreme northern and southern latitudes doesn't exactly invite scientists to spend protracted periods of research there; and thanks to their size and lifestyle, and possibly also as a result of their unspectacular appearance in comparison to orcas, they are rarely kept in dolphinariums, so they are not generally available for research in captivity either. Their affinity with bottle-nose dolphins and orcas, as well as their social behaviour and long nursing period of more than three years, suggest that they have attained a high state of cognitive development. This surmise is also confirmed by anatomical studies of their brains, which have revealed that around 37 billion neurons are present in their neocortex alone.[22] To put that figure in perspective, human beings have some 86 billion neurons in their entire brain.

The common history of humans and pilot whales is, like so many human–animal relationships, truly ambivalent. While pilot whales represent one of the chief tourist attractions for many thousands of whale watchers on the Canary Islands, the Faroese exploit the strong emotional bond between pilot whale family members, and with other species, and drive them all into shallow bays where they are slaughtered one by one. At one time it was not uncommon for the animals' death throes to last for several minutes, as the hunters concentrated their efforts on one whale, stabbing it repeatedly. During that time, the other pilot whales could see their family members, offspring, siblings and parents, along with other individuals they had formed an emotional attachment to over the course of their lives, being butchered. In Newfoundland alone between 1947 and 1971, more than 50,000 whales were killed in this way. On the Faroes, this practice began with human settlement of the islands some 1,200 years ago. At that time, hunting was conducted from robust, seagoing rowing boats. This epic struggle of human muscle power against the elements often literally ensured the survival of entire communities. If the pilot whale migrations failed to materialize, this meant a hard winter for the Faroese.

The modern inhabitants of the Faroe Islands are descendants of the Vikings and can be justifiably proud of their history, with their parliament, which goes back over 1,000 years, being one of the oldest legislatures in the world. Nowadays, the Faroes is an advanced welfare state with a living standard comparable to that of Switzerland or Norway.

For some fifty years now, the drive hunting of pilot whales on the Faroes has only been carried out for sport, and the islanders are no longer economically reliant on it.[23] Even so, the inhabitants regard this tradition as extremely important, as it strengthens a sense of cohesion within the community. The drive hunt has become a kind of folk festival, where people can join with their friends and relatives and work to a common purpose. Their task is made easier by modern boats and sophisticated technology.

In former times, the hunt would begin with fishermen banging all kinds of objects on the bottom of their boats to try to generate as much noise as possible. In addition, rocks attached to lines were thrown overboard to drive the whales ahead of the flotilla. Nowadays, things are a lot easier: you simply switch on echolocation equipment, and the disagreeable sound this makes herds the whales into one of twenty-three shallow bays on the islands reserved for the drive hunt. A successful hunt is cause for communal celebration, and the meat is distributed equitably and at no cost. The drive hunt therefore represents a cultural asset worth protecting in the community life of the Faroe Islands.

But for those who can't even bear to watch the gruesome videos on YouTube,[24] perhaps this brief description will suffice to convey the reality of what takes place: after the whales have been driven into knee-deep water, a group of men often several hundred strong swarms over them, armed with sharp metal hooks. These hooks are driven into the animals' bodies or their sensitive blowholes in order to drag them on land. If you've ever swallowed something the wrong way, you'll know how sensitive our windpipe is and may well be able to imagine how excruciatingly painful it must feel to be hauled from the water by the blowhole. Admittedly, this method produces less blood. The actual killing of the animals is carried out using a metal lance specially designed for that purpose and then by making a deep incision at the back of the whale's head.

For an experienced hunter, the whole process is a matter of just seconds' work. On the whales' part, though, the hunt and the stress and fear of death associated with it lasts considerably longer. An environmental journalist and friend of mine, Hans Peter Roth, has described this in the following graphic terms: 'Mothers lose their children and vice versa, while the stress causes other mother animals to give birth prematurely.' Meanwhile, on land, the following sight met his eyes: 'Children, girls and boys from the ages of four to fourteen were playing and jumping around the whales, with their boots all covered in blood and gore. To us it was a grotesque and deeply repellent spectacle.'

But as Roth,[25] who is also very familiar with the situation at Taiji, explained to me, there is one big difference between the Japanese and the Faroese. While the Japanese reacted towards him and the film crew that made *The Cove* partly with antipathy and partly also with physical violence, the Faroese actually invite interested journalists along and give them a blow-by-blow account of the hunt and the kill. This openness and friendliness only serves to underline that the Faroese have no sense that they're doing anything wrong. Consequently, children playing on the whale cadavers are just an expression for them of their country's custom being passed down the generations. Self-assuredly, and in some measure also in self-defence, they contrast inhumane factory-farming practices in the

EU with the animal husbandry in harmony with nature that they practise. The principal source of food for the Faroese is the sheep they keep, which roam free across the islands. And we would have to admit that the excruciating death by asphyxiation, lasting several minutes, that porpoises suffer as 'by-catch' in the gill nets of EU fishing boats is no less abhorrent.

Nevertheless, to anyone who has given even the most fleeting thought to the fact that pilot whales are self-aware, empathetic and systematic individuals with a conception of space and time, the bloody culture of the pilot whale drive hunt is a hateful atrocity. In my view, that is the real clincher: for as long as one remains ignorant of what one is doing or to whom one is doing it, even this kind of cruelty can be squared with your ethical and moral outlook.

VIII

Epilogue

L IKE MARTIN LUTHER, I grew up eating Thuringian bratwurst. In the absence of doner kebabs, sushi and the like, as
a kid I could sniff my way around the entire inner city of
Erfurt and pass a new booth grilling these delicious sausages every
couple of hundred metres. Those times are now long gone, yet the
Thuringian bratwurst is still highly prized in culinary circles, and
a few years ago I even bought a genuine Thuringian sausage from
a genuine Thuringian vendor on Wall Street in New York. The guy
must be a millionaire by now. Also, one of my fondest memories of
my late father is the secret raids we used to make together on our
larder at home, and the thick slices of salami we shared. In other
words, eating meat is part of my personal culture, and although
I'm now a vegetarian, I still like the taste of meat. Many of my
fellow Thuringians enjoy that taste, and so we occupy the top slot
of all the regions of Germany for meat consumption. Although,
at 85,000 tons annually, we produce comparatively more pork, we
also import 25 per cent of meat requirements.

After an absence of twenty years, a few years ago I moved back
to my home town of Erfurt. Even so, I still regularly experience
a culture shock whenever I go to a Thuringian supermarket. It's

often the case that the meat and sausage counter there is over 20 metres (65 feet) long. By comparison, the cheese counter rarely exceeds a measly couple of metres. A small army of bustling shop assistants are rushed off their feet serving customers, whereas at the cheese counter you easily have time to make a couple of calls on your mobile before a member of staff finally appears. It was exactly the opposite in Berlin.

Despite the strain that meat production is putting on the planet, and the warnings issued by the WHO regarding excessive meat consumption and people's growing awareness about aspects of animal welfare, over the last twenty years pork production has increased in Germany by 50 per cent, while the number of chickens processed has risen by almost 200 per cent. That's really odd, as the population of the country hasn't gone up in that time, but has remained stable at around 80 million for decades. Also, there are fewer meat-processing plants than there used to be; unfortunately, those that remain are growing larger. One of these huge plants is located in the idyllic municipality of Bad Kleinen on the northern shore of Lake Schwerin. According to 2016 official statistics,[1] the pig-fattening facility 'Tierzucht Gut Losten GmbH & Co. KG' houses around 34,000 pigs. You can easily gauge the sheer size of the place on Google Earth – I reckon it must cover around 100,000 square metres (120,000 square yards). That amounts to about 3 square metres (32 square feet) per animal. However, a pig can grow to a weight of about 200 kilograms (440 pounds), though it is usually taken to slaughter when it reaches half that size. The average European weighs about 80 kilos (175 pounds) – below the weight of a pig for slaughter – but has at his or her disposal (in Germany at least) a total living space of some 47 square metres (500 square feet). For the life of me, I cannot imagine how I would cope if I was not permitted to go beyond the bounds of my allotted 47 square metres for the whole of my life.

But now just imagine what it would be like if you couldn't leave your 3 square metres' worth of space for your entire life. Where German animal welfare laws are concerned, this method of rearing pigs does not cause them any harm. Otherwise the facility would not have been granted a licence to operate or would have been immediately closed down by the relevant government authorities. As it happens, German legislation on this score only requires an area of 2 square metres (21 square feet).[2]

Of course, scientific studies have been carried out on pig-rearing conditions, as farmers are well aware that stressed pigs yield less in the way of profit. They therefore have a vested interest in ensuring that their pigs are well treated. One method of telling if a pig is stressed or not was developed at the Leibniz Institute for Farm Animal Biology in Dummerstorf near Rostock. The researchers there invented a device that can tell from the grunts and other noises the pigs make whether they are under stress or not.[3] Now, I'm not involved in agribusiness, and I freely admit I don't know much about factory farming, but I still can't help but wonder what the employees of the pig-rearing plant do if one or more of the animals is found to be stressed. Do they go into the pens and comfort the pigs? Or do they have enough space to separate the pigs from one another now and do they then have sufficient time to reintroduce the pigs to one another in a stress-free environment? I really have no idea. But as a biologist I do know that using a stress detector will not only enable me to find out which animals are stressed – by implication, it will also tell me which animals aren't suffering from stress. As a breeder, I wouldn't find it hard to come to a commercially sensible decision and only select for rearing those animals that can cope better with my husbandry conditions. I'd hazard a guess that this form of selection isn't legal, since, according to German law, the conditions have to be suited to the animals and not the other way round. But who would be able and willing to monitor such a system?

Yet what goes on in a pig's brain, we might well ask? To answer this, let us first imagine a sow. We see her filth-encrusted snout and most likely her rear end is in a similar state. All in all, an animal that you'd rather approach wearing rubber boots. In 1965, a behavioural biology observation from Japan became famous around the world. Scientists there had discovered that a troop of indigenous macaques (*Macaca fuscata*) inhabiting Koshima Islet washed grubby sweet potatoes in a nearby stream.[4] Even today, this is still cited as a prime example of remarkable animal behaviour.

And it's true, such behaviour really is quite extraordinary, since it requires the animal to have developed a range of cognitive capacities. First, as a macaque I must be able to tell whether my food is dirty or clean. Second, I need to know that I can clean the food in water. Third, I have to deliberately carry the food from one location to another. And last but not least, I need to possess a certain amount of self-control, since I postpone the gratification of consuming the food to a later juncture.

We have considered all these examples already in the course of this book, and categorized them accordingly. But could you ever imagine pigs washing their food in a similar way if they were given the opportunity? Presumably not, and to be honest I'd also have been willing to bet that wasn't so. But if I had, I'd have lost my stake, since precisely such behaviour can on rare occasions be observed in the wild. The researchers who discovered this were so impressed by their observation that they attempted to replicate it in an experiment at Basel Zoo in Switzerland. The pigs there were given halved apples that had either been rubbed with sand or were unsullied. And sure enough, the pigs only opted to wash apples if they were dirty.[5] The common insult 'You dirty pig' therefore seems to hold no weight.

In a recently published overview article, the researchers compared the cognitive capacities of pigs with those of other animals

and of us humans. The findings brought to light a number of characteristics that one would not normally have associated with pigs.[6]

- They have a long-term memory.
- They possess spatial awareness.
- They can understand information in the form of symbols and
  have the ability to learn procedures expressed in this form.
- They have a pronounced propensity for play.
- They live in social communities with knowledge about other
  individuals.
- They possess the ability to learn from one another and to
  cooperate.
- They can operate a joystick in the same way primates do.
- They can use a reflection in a mirror to locate food.
- They exhibit empathy.

In an interview, the neuroscientist heading this experiment,
Loli Marino of Emory University, stated: 'We have shown that pigs
share a number of cognitive capacities with other highly intelligent
species such as dogs, chimpanzees, elephants, dolphins, and even
humans.'[7] To anyone who finds that too theoretical, I recommend
a really interesting video on YouTube, showing an experiment in
which six-week-old pigs were compared with eighteen-month-old
children.[8] Guess which group scored higher? Also, this is precisely
the age at which piglets in intensive rearing units are separated from
their mothers. In nature, they stay together for eighteen months.

In point of fact, cognitive research into pigs has only just begun.
At the Messerli Research Institute at the Veterinary University of
Vienna, a project was recently initiated to study the behaviour of
Kunekune pigs, a small breed of domestic pig from New Zealand.
In contrast to the generality of livestock research, which is all about
increasing yield, the purpose of this programme is to go back to

basics and look at the animal's natural behaviour. Perhaps its findings will enable us to understand how pigs should really be kept. The 2 square metres provided for by German legislators are frankly absurd, and officials in the relevant government departments must be perfectly well aware that under such housing conditions, the animals will suffer both physical and psychological damage, and that certainly isn't a legal state of affairs.

But what are we doing now that we have this knowledge? The answer is: absolutely nothing, of course! There's even a scientific term for this: cognitive dissonance. Let's stir up old Aesop one last time. In his fable 'The Fox and the Grapes', the fox turns up its nose and tells the grapes snootily: 'Oh, you aren't even ripe yet! I don't need any sour grapes.' Of course, the fox only said this because it wasn't able to reach the high-hanging bunch of grapes anyway. A good friend of mine has made this fable of Aesop into her life credo, and asks herself every day what she can convince herself she doesn't want, and that works just fine for her.

The only problem is that our perceptions, thoughts, feelings, opinions and desires are often not singing from the same hymn sheet. Our salary increase isn't in line with our expectations, or the political party we voted for hasn't kept its promises. This dissonance feels unpleasant, and we look for some way out of the dilemma.

Two examples: have you ever heard of the Benjamin Franklin effect? This shrewd politician realized that people whom we help become more agreeable to us. All we have to do is think of them as being more agreeable to us, because only then is our help an appropriate response. That also works the other way. In victim abasement, individuals to whom you are doing something terrible are divested of their humanity and also vilified in our minds. Victims of domestic abuse are called 'sluts' or 'wimps', victims of racism 'wogs' or 'Pakis', while victims of discrimination have traditionally been turned into slaves, and victims of animal cruelty are dubbed 'beasts'. They're just animals for slaughter, we tell ourselves, it's something we've always done, it's part of our culture and we have every right to act in this way.

We describe the world in a way that we would like it to be. Droves of social psychologists have covered and are still covering this topic, because it is one of the prime reasons why we humans so frequently fail to act in a rational and sensible way.

Let's look at this in detail. The technical literature talks about welfare for animals. In practice, this means avoiding causing pain, stress and suffering. Paragraph 2 of the German animal protection law states that 'the natural movements of an animal should not be restricted in such a way as to cause it pain or avoidable suffering or injury'. But even if this were the case in reality and a pig could actually move naturally in 2 square metres of space, would that constitute welfare? The well-known behavioural biologist Jonathan Balcombe had advocated shifting the focus to reward and pleasure because it is precisely those mechanisms that made development possible in the first place over the course of evolution. As we have seen in various chapters, these are also the same mechanisms that trigger the overwhelming majority of behaviours. In Balcombe's estimation, every living creature has a natural right to these stimuli.[9]

In our general sense of justice, only those individuals who are capable of bestowing rights themselves are deemed capable of having rights. If I demand rights of my own, then I must be able to abide by them myself.

So, if I want to ensure that no one steals my house or my banana, then it is incumbent upon me not to steal anyone else's house or banana. But am I entitled to steal something from someone who has no conception of property rights? As we learned in 'Death cults and war' and 'Self-awareness', there is a phenomenon called the endowment effect. The American economist Herbert Gintis believes that there is a compelling case for applying the concept of private property to animals, since the defensive behaviour displayed in safeguarding property developed as part of our evolutionary history and quite independently of our legal system.[10]

So, animals have rights. In reality, though, namely in our legal systems, they have none. Anyhow, we reason, they don't need any rights, since we have our protection legislation in place. Laws on animal welfare safeguard the animals in our care, and environmental protection laws protect animals in the wild. But if we consider that a pig that has been kept for its entire life in a space measuring 2 square metres suffers no ill effects, then surely our animal protection laws aren't fit for purpose, are they? Not long ago, I also highlighted a similarly absurd instance of environmental law concerning the protest lodged by the German environmental conservation NGO *Naturschutzbund Deutschland* (NABU) against development of the Butendiek offshore wind farm in the North Sea.[11] After almost twenty years working in environmental and animal conservation, I am now firmly of the opinion that our conservation laws are not fit for purpose. If conservation laws really had any teeth as a judicial mechanism, then you can be sure that corresponding ones would also have been framed to protect PLCs and limited liability companies. Instead of such laws, we have the concept of the legal person.

In her doctoral thesis, the legal scholar Caroline Raspé has put forward the hypothesis that we should, alongside our concepts of the natural person – namely ourselves – and the legal entity or person – namely companies, associations and charities – also introduce a third 'person' into jurisprudence, one that she calls the 'animal person'.[12] This 'animal person' could be represented by any lawyer. In her view, it doesn't matter whether the animal in question is a cognitively highly developed one, like a chimpanzee, or a snail.

Ultimately, in any lawsuit that arose, the lawyer should be able to adduce the rights of the animal, and these would necessarily turn on its capacities and entitlements.

For me, this point of view was a real eye opener. In my book 'Personal Rights for Animals',[13] I called for dolphins, great apes and elephants to be accorded the status of a person. This is a perfectly legitimate standpoint that has the support of several eminent jurists.[14] Unfortunately, I constantly got the uneasy feeling that my proposal might inadvertently create a two-class system for animals. Raspé's proposal that we add another person, namely an 'animal person', to our existing legal system is far more elegant. On this basis, we could define our relationship with animals afresh, and would have the tried-and-tested and flexible mechanisms of our justice system at our disposal.

A loose alliance of scientists met recently to lobby for this idea. This predominantly German-speaking group, which called itself the Individual Rights Initiative, includes almost every individual who has dealt with the topic of animal conservation in a scientific context. If you're interested in supporting its work, you can visit its website on www.iri.world.

While it can be presumed that many cognitively highly developed animals think and feel just like us on an individual level, they are not in a position to use force to resist our strategically planned and concertedly exercised abuse of them.

But wait a minute – it turns out that there *is* one such animal, and he certainly plans his behaviour strategically. The keepers in a zoo at Furuvik in Sweden have their work cut out with him, because although they regularly scour his enclosure for loose stones, no visitor to the zoo is truly safe around him. Over a number of years Santino, a male chimpanzee, has perfected his throwing technique and his stockpiling of stones, and carefully works out hiding places. He has also got to know the hollow sound made by the concrete walls of his enclosure as they gradually crumble and form cavities. This 'renewable resource' provides him with an excellent stock of new munitions.[15] And by the time he's exhausted his supply, the wall will have crumbled to nothing and freedom will beckon.

Endnotes

II: Going at it like animals

1   Hebert, P. D. N.: Tardigrada, *Encyclopedia of Earth* (2008), http://www. eoearth. org/view/article/156414.

2   Reinhardt, K., Siva-Jothy, M. T.: 'Biology of the bed bugs', *Annual Review of Entomology* (2007) 52, pp. 351–374.

3   Tautz, J., Heilman, H. R.: *Phänomen Honigbiene* (Wiesbaden, 2012).

4   Wallberg, A., Pirk, C. W., Allsopp, M. H., Webster, M. T.: 'Identification of multiple loci associated with social parasitism in honeybees', *PLoS Genetics* (2016) 12 (6).

5   Rijksen, H. B.: 'A field study on Sumatran orang utans (*Pongo pygmaeus abellii*) (Lesson 1827). PhD thesis, Nature Conservation Department, Agricultural University (Wageningen, 1978).

6   Thomsen, R., Sommer, V.: 'Masturbation (nonhuman primates)', *The International Encyclopedia of Human Sexuality* (2015).

7   www.youtube.com/watch?v=Gn64WPzw6_I.

8   www.youtube.com/watch?v=qVE6ozwXx1k.

9   www.hotdollfordog.com.

10   McGrew, W. C.: 'Chimpanzee technology', *Science* (2010) 147 (2).

11   www.nytimes.com/2010/05/04/science/04tier.html?_r=1.

12   Nishida, T.: 'The leaf-clipping display: a newly-discovered expressive gesture in wild chimpanzees', *Journal of Human Evolution* (1980) 9, pp. 117–128.

13   Bentley-Condit, V. K., Smith, E. O.: 'Animal tool use: current definitions and an updated comprehensive catalog', *Behaviour* (2010) 147 (2).

14   www.youtube.com/watch?v=cZU2YYYxEsw.

15   Boesch, C.: 'From material to symbolic cultures: culture in primates', *The Oxford Handbook of Culture and Psychology*, Oxford Library of Psychology (2012).

16   Cissewski, J., Boesch, C.: 'Communication without language – how great apes may cover crucial advantages of language without creating a system of symbolic communication', *Gesture* (2016) 15 (2), pp. 224–249.

17   Thornhill, R., Palmer, C.: *A Natural History of Rape: Biological Bases of Sexual Coercion* (Cambridge, 2001).

18   Schmotzer, B., Zimmerman, A.: 'Über die weiblichen Begattungsorgane der gefleckten Hyäne', *Anatomischer Anzeiger* (1922) 55, pp. 257–264.

19   Glickman, S. E., Zabel, C. J., Yoerg, S. I., Weldele, M. L., Drea, C. M., Frank, L. G.: 'Social facilitation, affiliation, and dominance in the social life of spotted hyenas', *Annals of the New York Academy of Sciences* (1997) 807, pp. 175–184.

20   www.spiegel.de/spiegel/print/d-122760764.html.

21   Izzo, T. J., Rodrigues, D. J., Menin, M., Lima, A. P., Magnusson, W. E.: 'Functional necrophilia: a profitable anuran reproductive strategy?', *Journal of Natural History* (2012) 46 (47–48).

22   Oelze, V. M., Fuller, B. T., Richards, M. P., Fruth, B., Surbeck, M., Hublin, J. J., Hohmann, G.: 'Exploring the contribution and significance of animal protein in the diet of bonobos by stable isotope ratio analysis of hair', *Proceedings of the National Academy of Sciences of the United States of America* (2011) 108 (24), pp. 9792–9797.

23   Manson, J. H., Perry, S., Parish, A. R.: 'Nonconceptive sexual behavior in bonobos and capuchins', *International Journal of Primatology* (1997) 18 (5), pp. 767–786.

24   De Waal, F. B. M.: 'Bonobo sex and society', *Scientific American Special Edition* (2006) 16 (2), pp. 14–21.

25   Surbeck, M., Deschner, T., Schubert, G., Weltring, A., Hohmann, G.: 'Mate competition, testosterone and intersexual relationships in bonobos (*Pan paniscus*)', *Animal Behaviour* (2012) 83 (3), pp. 659–669.

26   Connor, R. C., Watson-Capps, J. J., Sherwin, W. B., Krützen, M.: 'A new level of complexity in the male alliance networks of Indian Ocean bottlenose dolphins (*Tursiops* sp.)', *Biology Letters* (2011) 7 (4), pp. 623–626.

27   Cummins, F. S. et al.: 'Extreme aggression in male squid induced by a b-MSP-like pheromone', *Current Biology* (2011) 21 (4), pp. 322–327.

28   Buston, P.: 'Social hierarchies: size and growth modification in clownfish', *Nature* (2003) 424, p. 145 f.

29   Dunkel, L. P. et al.: 'Variation in developmental arrest among male orangu-

tans: a comparison between Sumatran and a Bornean population', *Frontiers in Zoology* (2013) 10.

30   Benenson, J. F., Tennyson, R., Wrangham, R. W.: 'Male more than female infants imitate propulsive motion', *Cognition* (2011) 121 (2), pp. 262–267.

31   Hassett, J. M., Siebert, E. R., Wallen, K.: 'Sex differences in rhesus monkey toy preferences parallel those of children', *Hormones and Behavior* (2008) 54 (3), pp. 359–364.

32   Kahlenberg, S. M., Wrangham, R. W.: 'Sex differences in chimpanzees' use of sticks as play objects resemble those of children', *Current Biology* (2010) 20, p. 1067 f.

33   Pinker, S.: *The Blank Slate: The Modern Denial of Human Nature* (New York, 2002).

34   Colapinto, J.: *As Nature Made Him: The Boy who was Raised as a Girl* (New York, 2000).

35   Bartels, A., Zeki, S.: 'The neural correlates of maternal and romantic love', *Neuro-Image* (2004) 21 (3), pp. 1155–1166.

36   Damasio, A.: 'Human behavior – brain trust', *Nature* (2005) 435, p. 571 f.

37   Lukas, D., Clutton-Brock, T. H.: 'The evolution of social monogamy in mammals', *Science* (2013) 341 (6145), pp. 526–530.

38   Burkett, J. P. et al.: 'Oxytocin-dependent consolation behavior in rodents', *Science* (2016) 351 (6271), pp. 375–378.

39   Kosfeld, M., Heinrichs, M., Zak, P. J., Fischbacher, U., Fehr, E.: 'Oxytocin increases trust in humans', *Nature* (2005) 435, pp. 673–676.

40   Reddon, A. R., O'Connor, C. M., Marsh-Rollo, S. E., Balshine, S.: 'Effects of isotocin on social responses in a cooperatively breeding fish', *Animal Behaviour* (2012) 84 (4), pp. 753–760.

41   Oliva, J. L., Rault, J. L., Appleton, B., Lill, A.: 'Oxytocin enhances the appropriate use of human social cues by the domestic dog (*Canis familiaris*) in an object choice task', *Animal Cognition* (2015) 18 (3), pp. 767–775.

42   Crews, D., Garstka, W.: 'The ecological physiology of a garter snake', *Scientific American* (1982) 247, pp. 159–168.

43   Piertney, S., Oliver, M.: 'The evolutionary ecology of the major histocompatibility complex', *Heredity* (2006) No. 96, pp. 7–21.

44   Woelfing, B., Traulsen, A., Milinski, M., Boehm, T.: 'Does intra-individual major histocompatibility complex diversity keep a golden mean?'. *Philosophical Transactions of the Royal Society* A (2009) 364 (1513).

45   Sommerfeld, R. D., Boehm, T., Milinski, M.: 'Desynchronising male and female reproductive seasonality: dynamics of male MHC-independent

olfactory attractiveness in sticklebacks', *Ethology Ecology & Evolution* (2008) 20 (4), pp. 325–336.

46   Pheromone parties are a dating fad that involves sniffing items of clothing whose owners remain anonymous. If you find the body odour appealing, it's the first step to finding your soulmate.

47   Rozin, P., Gruss, L., Berk, G.: 'The reversal of innate aversions: attempts to induce a preference for chili peppers in rats', *Journal of Comparative and Physiological Psychology* (1993) 79, pp. 1001–1014.

48   Kish, G. B., Donnenwerth G. V.: 'Sex differences in the correlates of stimulus seeking', *Journal of Consulting and Clinical Psychology* (1972) 38 (1), p. 42.

49   Byrnes, N. K., Hayes, J. E.: 'Behavioral measures of risk tasking, sensation seeking and sensitivity to reward may reflect different motivations for spicy food liking and consumption', *Appetite* (2016) 103, pp. 411–422.

III: Unknown cultures

1   Tylor, E. B.: *Anthropology* (London, 1881).

2   Rendell, L., Whitehead H.: 'Culture in whales and dolphins', *Behavioral and Brain Sciences* (2001) 24, pp. 309–382.

3   Culture is information or behaviour acquired from conspecifics through some form of social learning. Boyd, R., Richerson, P. J.: 'Why culture is common, but cultural evolution is rare', *Proceedings of the British Academy* (1996) 88, pp. 77–93.

4   Whiten, A., Goodall, J., McGrew, W. C., Nishida, T., Reynolds, V., Sugiyama, Y., Tutin, C. E., Wrangham, R. W., Boesch, C.: 'Chimpanzee cultures', *Nature* (1999) 399, pp. 682–685.

5   Laland, K. N., Galef, B. G.: *The question of animal culture* (Cambridge, 2009).

6   Laland, K. N., Janik, V. M.: 'The animal cultures debate', *Trends in Ecology and Evolution* (2006) 21 (10).

7   Krützen, M., van Schaik, C., Whiten, A.: 'Response to Laland and Janik: The animal cultures debate', *Trends in Ecology and Evolution* (2006) 22 (1).

8   Huber, S. K.: 'Reproductive isolation of sympatric morphs in a population of Darwin's finches', *Proceedings of the Royal Society* (2007) 274, (D 709–1714).

9   Esposito, G. et al.: 'Infant calming responses during maternal carrying in humans and mice', *Current Biology* (2013) 23 (9), pp. 739–745.

10   Warner, R. R. 'Traditionality of mating-site preferences in a coral reef fish', *Nature* (1988) 335, pp. 719–721.

11   McGrew, W. C.: *The cultured chimpanzee* (Cambridge, 2004).

12   Whiten, A., Schaik, C. P.: 'The evolution of animal "cultures" and social intelligence', *Philosophical Transactions of the Royal Society* (2007) 362, pp. 603–620.

13   Haidle, M. N., Conard, N. J., Bolus, M. (eds.): *The Nature of Culture*. Book publication based on an Interdisciplinary Symposium The Nature of Culture, Tübingen. Vertebrate Paleobiology and Paleoanthropology (2016).

14   www.youtube.com/watch?v=RxFCIXEAf8c.

15   au.youtube.com/watch?v=OlB70VP8MPY.

16   Payne, K., Payne, R.: 'Large-scale changes over 19 years in songs of humpback whales in Bermuda', *Zeitschrift für Tierpsychologie* (1985) 68, pp. 89–114.

17   Noad, M. J. et al.: 'Cultural revolution in whale song', *Nature* (2001) 408, p. 537.

18   Garland, E. C. et al.: 'Dynamic horizontal cultural transmission of humpback whale song at the ocean basin scale', *Current Biology* (2011) 21 (8), pp. 687–691.

19   Riesch, R., Barrett-Lennard, L. G., Ellis, G. M., Ford, J. B., Deecke, V. B.: 'Cultural traditions and the evolution of reproductive isolation: ecological speciation in killer whales?', *Biological Journal of the Linnean Society* (2012) 106, pp. 1–17.

20   Rendell, L. E., Whitehead, H.: 'Culture in whales and dolphins', *Behavioral and Brain Sciences* (2001) 24, pp. 309–324. See also Lennard, L. G., Deecke V. B., Yurk H., Ford J. K. B.: 'A sound approach to the study of culture', *Behavioral and Brain Sciences* (2001) 24, pp. 325–326.

21   Riesch, R., Barrett-Lennard, L. G., Ellis, G. M., Ford, J. B., Deecke, V. B.: 'Cultural traditions and the evolution of reproductive isolation: ecological speciation in killer whales?', *Biological Journal of the Linnean Society* (2012) 106, pp. 1–17.

22   Richerson, P. J., Boyd, R.: *Not by Genes Alone: How Culture Transformed Human Evolution* (Chicago, 2005).

23   Ford, J. K. B., Ellis, G. M.: *Transients – Mammal-hunting Killer Whales* (Vancouver, 1999).

24   Matkin, C. O., Saulitis, E. L., Ellis, G. M., Olesiuk, P., Rice, S. D.: 'Ongoing population-level impacts on killer whales *Orcinus orca* following the "Exxon Valdez" oil spill in Prince William Sound, Alaska', *Marine Ecology Progress Series* (2008) 356, pp. 269–281.

25   Milius, S.: 'Getting the gull: baiting trick spreads among killer whales. (*Orcinus orca*)', *Science News* (2005) 168 (8), p. 118.

26   Semaw, S.: 'The world's oldest stone artefacts from Gona, Ethiopia: their implications for understanding stone technology and patterns of human evolution between 2.6–1.5 million years ago', *Journal of Archaeological Science* (2000) 27, pp. 1197–1214.

27   Sauciuc, G. A., Persson, T., Bååth, R., Bobrowicz, K., Osvath, M.: 'Affective forecasting in an orang-utan: predicting the hedonic outcome of novel juice mixes', *Animal Cognition* (2016) 19 (6), pp. 1081–1092.

28   www.ted.com/talks/luis_von_ahn_massive_scale_online_collaboration#t-128360.

29   Luncz, L. V., Mundry, R., Boesch, C.: 'Evidence for cultural differences between neighboring chimpanzee communities', *Current Biology* (2012) 22 (10), pp. 922–926.

30   Mercader, J., Barton, H., Gillespie, J., Harris, J., Kuhn, S., Tyler, R., Boesch, C.: '4,300-Year-old chimpanzee sites and the origins of percussive stone technology', *Proceedings of the National Academy of Sciences* (2007) 104 (9) pp. 3043–3048.

31   Whiten, A., Spiteri, A., Horner, V., Bonnie, K. E., Lambeth, S. P., Schapiro, S. J., de Waal, F. B. M.: 'Transmission of multiple traditions within and between chimpanzee groups', *Current Biology* (2007) 17 (12), pp. 1038–1043.

32   www.youtube.com/watch?v=ScqG54B4KtE.

33   Sommer, V., Buba, U., Jesus, G., Pascual-Garrido, A.: 'Sustained myrme-cophagy in Nigerian chimpanzees: preferred or fallback food?', *American Journal of Physical Anthropology* (2017) 162 (2), pp. 328–336.

34   Humle, T., Matsuzawa, T.: 'Ant-dipping among the chimpanzees of Bossou, Guinea, and some comparisons with other sites', *American Journal of Primatology* (2002) 58, pp. 133–148.

35   Reindl, E., Beck, S. R., Apperly, I. A., Tennie, C.: 'Young children spontaneously invent wild great apes' tool-use behaviours', *Proceedings of the Royal Society B* (2016) 283 (1825).

36   Fishlock, V., Caldwell, C., Lee, P. C.: 'Elephant resource-use traditions', *Animal Cognition* (2016) 19, pp. 429–433.

37   Hart, B. L., Hart, L. A., McCoy, M., Sarath, C. R.: 'Cognitive behaviour in Asian elephants: use and modification of branches for fly switching', *Animal Behaviour* (2001) 62, pp. 839–847.

38   Nihei, Y., Higuchi, H.: 'When and where did crows learn to use automobiles as nutcrackers?', *Tohoku Psychologica Folia* (2001) 60, pp. 93–97.

39   Aplin, M. L., Farine, D. R., Morand-Ferron, J., Cockburn, A., Thornton, A., Sheldon, B. C.: 'Experimentally induced innovations lead to persistent culture via conformity in wild birds', *Nature* (2015) 518 (7540) pp. 538–541.

A video of the analysis of these social networks can be downloaded: www.ncbi.nlm.nih.gov/pmc/articles/PMC4344839/bin/NIHMS60796-supplement-video3.mp4.

40 Smolker, R. A.: 'Sponge carrying: a puzzle in the behavior of bottlenose dolphins' in: Seventh Biennial Conference on the Biology of Marine Mammals (1987) 5–9 December, Miami, Florida, p. 65.

41 Smolker, R. A., Richards, A., Connor, R. C., Mann, J., Berggren, P.: 'Sponge carrying by dolphins (Delphinidae, *Tursiops* sp.): a foraging specialization involving tool use?', *Ethology* (1997) 103 (6), pp. 454–465.

42 Krützen, M., Mann, J., Heithaus, M.R., Connor, C., Bejder, L., Sherwin, W. B.: 'Cultural transmission of tool use in bottlenose dolphins', *Proceedings of the National Academy of Sciences* (2005) 102, pp. 8939–8943.

43 Allen, S. J., Bejder, L., Krützen, M.: 'Why do Indo-Pacific bottlenose dolphins (*Tursiops* sp.) carry conch shells (*Turbinella* sp.) in Shark Bay, Western Australia?', *Marine Mammal Science* (2011) 27, pp. 449–454.

44 Endler, J. A., Endler, L. C., Doerr, N. R.: 'Great bowerbirds create theaters with forced perspective when seen by their audience', *Current Biology* (2010) 20, pp. 1679–1684.

45 Bravery, B. D., Nicholls, J. A., Goldize, A. W.: 'Patterns of painting in satin bowerbirds *Ptilonorhynchus violaceus* and males' responses to changes in their paint', *Journal of Avian Biology* (2006) 37, pp. 77–83.

46 Chisholm, A. H.: 'The use by birds of 'tools' or 'instruments'. *Ibis* (1954) 96 (3), pp. 380–383.

47 Neville, B.: 'The strange case of Billy Bowerbird', *Geo* (1988) 10, pp. 73–79.

48 Madden, J. R.: 'Do bowerbirds exhibit cultures?', *Animal Cognition* (2008) 11, pp. 1–12.

49 Bernardi, G.: 'The use of tools by wrasses (Labridae)', *Coral Reefs* (2011) 31 (1), p. 39.

50 www.youtube.com/watch?v=P_MYQy_eeTQ&feature=youtube.

51 Brockmann, H. J.: 'Tool use in digger wasps (Hymenoptera: Sphecinae)', *Psyche* (1985) 92, pp. 309–330.

52 Banschbach, V. S., Brunelle, A., Bartlett, K. M., Grivetti, J. Y., Yeamans, R. L.: 'Tool use by the forest ant *Aphaenogaster rudis*: ecology and task allocation', *Insectes Sociaux* (2006) 53 (4), pp. 463–471.

53 Henry, P. Y., Aznar, C.: 'Tool-use in Charadrii: active bait-fishing by a herring gull', *Waterbirds* (2006) 29 (2), p. 233 f.

54 Robinson, S. K.: 'Use of bait and lures by green-backed herons in Amazonian Peru', *The Wilson Bulletin* (1994) 106 (3), p. 567 ff.

55 www.youtube.com/watch?v=y_8hPcnGeCI.

56   Levey, D. J., Duncan R. S., Levins C. F.: 'Use of dung as a tool by burrowing owls', *Nature* (2004) 431, p. 39.

57   Marshall, M.: 'Alligators use tools to lure in bird prey', *New Scientist* (2013) 220 (2948–2949), p. 16.

58   www.youtube.com/watch?v=G32YehcdUAw.

59   Matrosova, V. A., Schneiderová, I., Volodin, I. A., Volodina, E. V.: 'Species-specific and shared features in vocal repertoires of three Eurasian ground squirrels (genus *Spermophilus*)', *Acta Theriologica* (2012) 57 (1), pp. 65–78. See also Slobodchikoff, C. N., Kiriazis, J., Fischer, C., Creef, E.: 'Semantic information distinguishing individual predators in the alarm calls of Gunnison's prairie dogs', *Animal Behaviour* (1991) 42, pp. 713–719.

60   Molnár, C., Pongrácz, P., Faragó, T., Dóka, A., Miklósi, A.: 'Dogs discriminate between barks: The effect of context and identity of the caller', *Behavioural Processes* (2009) 82, pp. 198–201.

61   Enggist-Dueblin, P., Pfister, U.: 'Cultural transmission of vocalizations in ravens, *Corvus corax*', *Animal Behaviour* (2002) 64, pp. 831–841.

62   Griesser, M.: 'Mobbing calls signal predator category in a kin group-living bird species', *Proceedings. Biological Sciences* (2009) 276 (1669), pp. 2887–2892.

63   Ratnayake, C. P., Goodale, E., Kotagama, S. W.: 'Two sympatric species of passerine birds imitate the same raptor calls in alarm contexts', *Naturwissenschaften* (2010) 97, pp. 103–108.

64   Stoeger, A. S., Mietchen, D., Oh, S., de Silva, S., Herbst, C. T., Kwon, S., Fitch, W. T.: 'An asian elephant imitates human speech', *Current Biology* (2012) 22 (22), pp. 2144–2148.

65   Ridgway, S., Carder, D., Jeffries, M., Todd, M.: 'Spontaneous human speech mimicry by a cetacean', *Current Biology* (2012) 22 (20), pp. 860–861.

66   www.youtube.com/watch?v=K4Uy_QOQfbs.

67   auditoryneuroscience.com/vocalizations-speech/hoover-talking-seal.

68   Fitch, W. T., Jarvis, E. D.: 'Birdsong and other animal models for human speech, song, and vocal learning', *Language, Music and the Brain* (Cambridge, 2012), pp. 499–540.

69   Pepperberg, I. M.: 'Evolution of Communication from an Avian Perspective' in D. Kimbrough Oller and Ulrike Griebel, editors, *Evolution of Communication Systems: A Comparative Approach* (Cambridge, 2004), pp. 171–192.

70   Pepperberg, I. M., Gordon, J. D.: 'Numerical comprehension by a grey parrot (*Psittacus erithacus*), including a zero-like concept', *Journal of Comparative Psychology* (2005) 119, pp. 197–209.

71 www.youtube.com/watch?v=utkb1nOJnD4.

72 Laland, K., Wilkins, C., Clayton, N.: 'The evolution of dance', *Current Biology* (2016) 26 (1), pp. 5–9.

73 Falk, D.: 'Comparative anatomy of the larynx in man and the chimpanzee: implications for language in Neanderthal', *American Journal of Physical Anthropology* (1975) 43, pp. 123–132.

74 www.johnclilly.com.

75 Herzing, D. L., Delfour, F., Pack, A. A.: 'Responses of human-habituated wild Atlantic spotted dolphins to play behaviors using a two-way human/ dolphin interface', *International Journal of Comparative Psychology* (2012) 25, pp. 137–165.

76 play.google.com/store/apps/details?id=com.shazam.android&hl=de.

77 www.wilddolphinproject.org/chat-is-it-a-dolphin-translator-or-an-interface.

78 www.ted.com/talks/denise_herzing_could_we_speak_the_language_of_ dolphins.

79 Herman, L. M., Richards, D. G., Wolz, J. P.: 'Comprehension of sentences by bottlenosed dolphins', *Cognition* (1984) 16, pp. 129–219.

80 Herman, L. M., Morrel-Samuels, P., Pack, A.: 'Bottlenosed dolphin and human recognition of veridical and degraded video displays of an artificial gestural language', *Journal of Experimental Psychology: General* (1990) 119, pp. 215–230.

81 Herman, L. M., Kuczaj, S., Holder, M. D.: 'Responses to anomalous gestural sequences by a language-trained dolphin: evidence for processing of semantic relations and syntactic information', *Journal of Experimental Psychology: General* (1993) 122, pp. 184–194.

82 Reiss, D., McCowan, B.: 'Spontaneous vocal mimicry and production by bottlenose dolphins (*Tursiops truncatus*): evidence for vocal learning', *Journal of Comparative Psychology* (1993) 107, pp. 301–312.

83 Brensing, K., Linke, K., Todt, D.: 'Sound source location by phase differences of signals', *Journal of the Acoustical Society of America* (2001) 109, pp. 430–433.

84 Herzing, D. L.: 'Clicks, whistles and pulses: passive and active signal use in dolphin communication', *Acta Astronautica* (2014) 105 (2), pp. 534–537.

85 Bradbury, J. W., Balsby, T. J. S.: 'The functions of vocal learning in parrots', *Behavioral Ecology and Sociobiology* (2016) 70 (3), pp. 293–312.

86 Suzuki, T. N., Wheatcroft, D., Griesser, M.: 'Experimental evidence for compositional syntax in bird calls', *Nature Communications* (2016) 7, No. 10986.

87   Ford, J. K. B.: 'Vocal traditions among resident killer whales (*Orcinus orca*) in coastal waters of British Columbia', *Canadian Journal of Zoology* (1991) 69, pp. 1454–1483.

88   Deeke, V. B.: 'Stability and change of killer whale (*Orcinus orca*) dialects', Thesis, University of British Columbia Library 1998.

89   Filatova, O. A., Fedutin, I. D., Burdin, A. M., Hoyt, E.: 'The structure of the discrete call repertoire of killer whales *Orcinus orca* from Southeast Kamchatka', *Bioacoustics* (2007) 16, pp. 261–280.

90   Deecke, V. B., Slater, Ü. J. B., Ford, J. K. B.: 'Selective habituation shapes acoustic predator recognition in harbour seals', *Nature* (2002) 420, pp. 171ff.

91   Foote, A.D., Griffin, R. M., Howitt, D., Larsson, L., Miller, P. J. O., Hoelzel, A. R.: 'Killer whales are capable of vocal learning', *Biology Letters* (2006) 2 (4), pp. 509–512.

92   Musser, W. B., Bowles, A. E., Grebner, D. M., Crance, J. L.: 'Differences in acoustic features of vocalizations produced by killer whales cross-socialized with bottlenose dolphins', *Journal of the Acoustical Society of America* (2014) 136 (4).

93   Hausberger, M., Bigot, E., Clergeau, P.: 'Dialect use in large assemblies: a study in European starling *Sturnus vulgaris* roosts', *Journal of Avian Biology* (2008) 39 (6), pp. 672–682.

94   Enggist-Dueblin, P., Pfister, U.: 'Cultural transmission of vocalizations in ravens, *Corvus corax*', *Animal Behaviour* (2002) 64, pp. 831–841.

95   Chen, Y. et al.: '"Compromise" in echolocation calls between different colonies of the intermediate leaf-nosed bat (*Hipposideros larvatus*)', *Public Library of Science One* (2016) 11 (3).

96   Melendez, K. V., Feng, A. S.: 'Communication calls of little brown bats display individual-specific characteristics', *Journal of the Acoustical Society of America* (2010) 128 (2), pp. 919–923.

97   Crockford, C., Herbinger, I., Vigilant, L., Boesch, C.: 'Wild chimpanzees produce group-specific calls: a case for vocal learning?', *Ethology* (2004) 10, pp. 221–243.

98   www.youtube.com/watch?v=3KJzDhMfWW8.

99   Arriaga, G., Zhou, E. P., Jarvis, E. D.: 'Of mice, birds, and men: the mouse ultrasonic song system has some features similar to humans and song-learning birds', *Public Library of Science One* (2012) 7 (10).

100 Hoier, S., Pfeifle, C., von Merten, S., Linnenbrink, M.: 'Communication at the garden fence – context dependent vocalization in female house mice', *Public Library of Science One* (2016) 11(3).

101 Personal correspondence with Ms. Christine Pfeifle, MPI Plön Leitung Maushaltung.

102 Oostenbroek, J. et al.: 'Comprehensive longitudinal study challenges the existence of neonatal imitation in humans', *Current Biology* (2016) 26(10), pp. 1334–1338.

103 Hobaiter, C., Byrne, R. W.: 'The gestural repertoire of the wild chimpanzee', *Animal Cognition* (2011) 14, pp. 745–767.

104 Byrne, R. W., Cochet, H.: 'Where have all the (ape) gestures gone?', *Psychonomic Bulletin & Review* (2016) 24 (1), pp. 68–71.

105 Douglas, P. H., Moscovice, L. R.: 'Pointing and pantomime in wild apes? Female bonobos use referential and iconic gestures to request genito-genital rubbing', *Scientific Reports* (2015) 5, No. 13999.

106 Gardner, R. A., Gardner, B. T., van Cantfort, T. E.: *Teaching Sign Language to Chimpanzees* (Albany, 1989).

107 Gardner, R. A., Gardner, B. T.: 'Comparative psychology and language acquisition' *Annals of the New York Academy of Sciences* (1978) 309, Psychology: The State of the Art, pp. 37–76.

108 www.nytimes.com/2007/11/01/science/01chimp.html?_r=2.

109 Fouts, R. S.: Language: 'Origins, definition, and chimpanzees', *Journal of Human Evolution* (1974) 3, pp. 475–482.

110 Rumbaugh, D. M.: *Language Learning by a Chimpanzee* (New York, 1977).

111 Savage-Rumbaugh, E. S., McDonald, E., Sevcik, R. A., Hopkins, W. D., Rupert, E.: 'Spontaneous symbol acquisition and communicative use by pygmy chimpanzees [*Pan paniscus*]', *Journal of Experimental Psychology* (1986) 112, pp. 211–235.

112 Savage-Rumbaugh, S., Lewin, R.: Kanzi. *The Ape at the Brink of the Human Mind* (Hoboken, 1994).

113 Clarke, E., Reichard, U. H., Zuberbühler, K.: 'The syntax and meaning of wild gibbon songs', *Public Library of Science One* (2006).

114 www.youtube.com/watch?v=JLOn8Fop96s.

115 *Brandt, R.: Können Tiere denken? Ein Beitrag zur Tierphilosophie* (Frankfurt/M, 2009).

116 Hare, B., Tomasello, M.: 'Domestic dogs (*Canis familiaris*) use human and conspecific social cues to locate hidden food', *Journal of Comparative Psychology* (1999) 113, pp. 173–177.

117 Mangelsdorf, M.: 'Interaktion zwischen Menschen und Pferden – im Vergleich zu Wölfen und Hunden in Interspezies-Kommunikation. Voraussetzungen und Grenzen', *PeriLog Freiburger Beiträge zur Kultur- und Sozialforschung* (2014) 7, pp. 42–64.

118   Malavasi, R., Huber, L.: 'Evidence of heterospecific referential communication from domestic horses (*Equus caballus*) to humans', *Animal Cognition* (2016) 19 (5).

119   Tschudin, A., Call, J., Dunbar, R. I. M., Harris, G., van der Elst, C.: 'Comprehension of signs by dolphins (*Tursiops truncatus*)', *Journal of Comparative Psychology* (2001) 115, pp. 100–105.

120   Xitco Jr., M. J., Gory, J. D., Kuczaj, S. A.: 'Dolphin pointing is linked to the attentional behavior of a receiver', *Animal Cognition* (2004) 7 (4), pp. 231–238.

121   Bräuer, J., Call, J., Tomasello, M.: 'All great ape species follow gaze to distant locations and around barriers', *Journal of Comparative Psychology* (2005) 119, pp. 145–154.

122   Leavens, D. A. Hopkins, W. D.: 'Intentional communication by chimpanzees (*Pan troglodytes*): a cross-sectional study of the use of referential gestures', *Developmental Psychology* (1998) 34, pp. 813–822.

123   Pika, S., Bugnyar, T.: 'The use of referential gestures in ravens (*Corvus corax*) in the wild', *Nature Communications* (2011) 2 (560).

124   Tomasello, M.: 'Why don't apes point?' in Endfield, N., Levinson, S., editors: *Roots of Human Sociality* (Oxford, 2006).

125   Scott-Phillips, T. C.: 'Meaning in animal and human communication', *Animal Cognition* (2015) 18 (3), pp. 801–805.

126   Moore, R.: 'Meaning and ostension in great ape gestural communication', *Animal Cognition* (2015) 19 (1).

127   www.cms.int/sites/default/files/document/COP11_Doc_23_2_4_Conservation_ Implications_Cetacean_En.pdf.

128   McComb, K., Moss, C., Durant, S. M., Baker, L., Sayialel, S.: 'Matriarchs as repositories of social knowledge in African elephants', *Science* (2001) 292, pp. 491–494.

129   Bradshaw, G. A., Schore, A. N., Brown, J. L., Poole, J. H., Moss, C. J.: 'Elephant breakdown social trauma: early disruption of attachment can affect the physiology, behaviour and culture of animals and humans over generations', *Nature* (2005) 433.

130   Ford, J. K. B., Ellis, G. M., Balcomb, K. C.: *Killer Whales: The Natural History and Genealogy of Orcinus orca in British Columbia and Washington State* (Vancouver, 2000). See also Barrett-Lennard, L. G., Deecke, V. B., Yurk, H., Ford, J. K. B.: 'A sound approach to the study of culture', *Behavioral & Brain Sciences* (2001) 24, pp. 325 f.

131   Teaney, D. O.: 'The insignificant killer whale: a case study of inherent flaws in the wildlife services' distinct population segment policy and a proposed solution', *Environmental Law* (2004) 34, pp. 647–702.

132 Ryan, S. J.: 'The role of culture in conservation planning for small or endangered populations', *Conservation Biology* (2006) 20, pp. 1321–1324.

133 Ibid.

IV: A Sense of Community

1   Propeller propulsion in bacteria: The flagella consist of long, spiral-shaped strands of protein that are some 15 to 20 nm thick. They rotate at a frequency of around 40–50 Hz and more. The fastest identified so far are two species of archaean (*Methanocaldococcus jannaschii* and *Methanocaldococcus villosus*) with speeds of up to 400 to 500 bps (body-lengths per second). A sportscar with 400 bps would travel at 6,000 km/h). idw-online.de/de/news465303.

2   *Mikrobiologischer Lehrpfad*, Max Planck Institut für terrestrische Mikrobiologie, Lehrtafel: Bakterien mit Gemeinschaftssinn.

3   www.aberwitzig.com.

4   North, G.: 'The biology of fun and the fun of biology', *Current Biology* (2015) 25 (1), pp. R1–R2.

5   www.youtube.com/watch?v=3dWw9GLcOeA.

6   Emery, N. J., Clayton, N. S.: 'Do birds have the capacity for fun?', *Current Biology* (2015) 25 (1), pp. R16–R20.

7   www.youtube.com/watch?v=bsiqdl6vsGQ.

8   Berridge, K. C., Kringelbach, M. L.: 'Affective neuroscience of pleasure: reward in humans and animals', *Psychopharmacology* (2008) 199, pp. 457–480.

9   Meijer, J. H., Robbers, Y.: 'Wheel running in the wild', *Proceedings of the Royal Society B* (2014) 281 (1786).

10 Riters, L. V.: 'Pleasure seeking and birdsong', *Neuroscience & Biobehavioral* (2011) 35, pp. 1837–1845.

11 www.youtube.com/watch?v=_mOyzDCC8ww.

12 www.youtube.com/watch?v=xHHIABb_qP4.

13 Burghardt, G. M.: 'A brief glimpse at the long evolutionary history of play', *Animal Behavior and Cognition* (2014) 1, pp. 90–98.

14 Burghardt, G. M., Dinets, V., Murphy, J. B.: 'Highly repetitive object play in a cichlid fish (*Tropheus duboisi*)', *Ethology* (2015) 121 (1), pp. 38–44.

15 Burghardt, G. M.: 'Play in fishes, frogs and reptiles', *Current Biology* (2015) 25 (1).

16 www.youtube.com/watch?v=oGvMAg25sVA.

17   www.youtube.com/watch?v=p-zGIS-WWZQ.

18   Zylinski, S.: 'Fun and play in invertebrates', *Current Biology* (2015), 25 (1).

19   Dapporto, L., Turillazzi, S., Palagi, E.: 'Dominance interactions in young adult paper wasp (*Polistes dominulus*) foundresses: a playlike behavior?', *Journal of Comparative Psychology* (2006) 120, pp. 394–400.

20   Pruitt, J. N., Burghardt, G. M., Riechert, S. E.: 'Non-conceptive sexual behavior in spiders: a form of play associated with body condition, personality type, and male intrasexual selection', *Ethology* (2012) 118, pp. 33–40.

21   Meijer, J. H., Robbers, Y.: 'Wheel running in the wild', *Proceedings of the Royal Society B* (2014) 281 (1786).

22   Bekoff, M.: 'Play signals as punctuation: the structure of social play in canids', *Behaviour* (1995) 132, pp. 419–429.

23   Bekoff, M.: 'Playful fun in dogs', *Current Biology* (2015) 25 (1).

24   Thalmann, O. et al.: 'Complete mitochondrial genomes of ancient canids suggest a European origin of domestic dogs', *Science* (2013) 342 (6160), pp. 871–874.

25   Blumstein, D. T., Chung, L. K., Smith, J. E.: 'Early play may predict later dominance relationships in yellow-bellied marmots (*Marmota flaviventris*)', *Proceedings of the Royal Society B* (2013) 280 (1759).

26   Behncke, I.: 'Play in the Peter Pan ape', *Current Biology* (2015) 25 (1), p. R24.

27   www.youtube.com/playlist?list=PLipu9sylv7l75dT3sFJXQa174M7bwfEvg.

28   De Waal, F. B. M.: 'Bonobo sex and society', *Scientific American Special Edition* (2006) 16 (2), pp. 14–21.

29   www.youtube.com/watch?v=rNWPqfCJDnc.

30   Ross, M. D., Owren, M. J., Zimmermann, E.: 'Reconstructing the evolution of laughter in great apes and humans', *Current Biology* (2009) 19 (13), pp. 1106–1111.

31   www.youtube.com/watch?v=jdzi8JFxoys.

32   Surbeck, M., Mundry, R., Hohmann, G.: 'Mothers matter! Maternal support, dominance status and mating success in male bonobos (*Pan paniscus*)', *Proceedings of the Royal Society B* (2011) 278 (1705).

33   Emma, A. et al.: 'Adaptive prolonged postreproductive life span in killer whales', *Science* (2012) 337 (6100), p. 1313.

34   Kruuk, H: *The Spotted Hyena: A Study of Predation and Social Behaviour* (Berkeley, 1972).

35   Macdonald, D.: *The Velvet Claw: A Natural History of the Carnivores* (New York, 1992).

36   Lewin, N., Treidel, L. A., Holekamp, K. E., Place, N. J.,Haussmann, M. F.:

'Socioecological variables predict telomere length in wild spotted hyenas', *Biology Letters* (2015) 11 (2).

37   Strandburg-Peshkin, A., Farine, D., Couzin, I., Crofoot, M. C.: 'Shared decision-making drives collective movement in wild baboons', *Science* (2015) 348 (6241), pp. 1358–1361.

38   Haun, D. B. M., Rekers, Y., Tomasello, M.: 'Majority-biased transmission in chimpanzees and human children, but not orangutans', *Current Biology* (2012) 22 (8).

39   Kulik, L., Langos, D., Widdig, A.: 'Mothers make a difference: mothers develop weaker bonds with immature sons than daughters', *Public Library of Science One* (2016) 11 (5).

40   CITES is the acronym for the Convention on International Trade in Endangered Species of Wild Fauna and Flora. www.cites.org.

41   Loftus, E.: 'Creating false memories', *Scientific American* (1997) 277 (3), pp. 70–75.

42   faculty.washington.edu/eloftus/.

43   www.ted.com/talks/elizabeth_loftus_the_fiction_of_memory#t-1034451.

44   Cheke, L. G., Simons, J. S., Clayton, N. S.: 'Higher body mass index is associated with episodic memory deficits in young adults', *Quarterly Journal of Experimental Psychology* (2016) 69 (11).

45   Roy, D. S. et al.: 'Memory retrieval by activating engram cells in mouse models of early Alzheimer's disease', *Nature* (2016) 531, pp. 508–512.

46   Ramirez, S. et al.: 'Creating a false memory in the hippocampus', *Science* (2013) 341 (6144), pp. 387–391.

47   science.sciencemag.org/highwire/filestream/594601/field_highwire_adjunct_files/0/Ramirez-SM.pdf.

48   Byrne, R. W., Bates, L. A., Moss, C. J.: 'Elephant cognition in primate perspective', *Comparative Cognition & Behavior Reviews* (2009) 4, pp. 1–15.

49   Foley, C. A. H., Pettorelli, N., Foley, L.: 'Severe drought and calf survival in elephants', *Biology Letters* (2008) 4, pp. 541–544.

50   Bruck, J. N.: 'Decades-long social memory in bottlenose dolphins', *Proceedings of the Royal Society* (2013) 280 (1768).

51   Clayton, N. S., Russell, J., Dickinson, A.: 'Are animals stuck in time or are they chronesthetic creatures?', *Topics in Cognitive Science* (2009) 1, pp. 59–71.

52   Zhang, S., Schwarz, S., Pahl, M., Zhu, H., Tautz, J.: 'Honeybee memory: a honeybee knows what to do and when', *Journal of Experimental Biology* (2006) 209 (22) pp. 4420–4428.

53   Hamilton, W. D.: 'Geometry of the selfish herd', *Journal of Theoretical Biology* (1971) 31 (2), pp. 295–311.

54   van Schaik, C. P.: 'The socioecology of fission-fusion sociality in orangutans', *Biomedical and Life Sciences* (1999) 40 (1), S: 69–86.

55   Archie, E. A., Moss, C. J., Alberts, S. C.: 'The ties that bind: genetic relatedness predicts the fission and fusion of social groups in wild African elephants', *Proceedings of the Royal Society B* (2005) 273, pp. 513–522.

56   Meggan, E. C., Volz, E., Packer, C., Ancel Meyers, L.: 'Disease transmission in territorial populations: the small-world network of Serengeti lions', *Journal of The Royal Society Interface* (2011) 8, pp. 776–786.

57   Smith, J. E., Sandra, K. M., Kay, E. H.: 'Rank-related partner choice in the fission-fusion society of the spotted hyena (*Crocuta crocuta*)', *Behavioral Ecology and Sociobiology* (2007) 61 (5), pp. 753–765.

58   Albon, S. D., Staines, H. J., Guinness, F. E., Clutton-Brock, T. H.: 'Density-dependent changes in the spacing behaviour of female kin in red deer', *Journal of Animal Ecology* (1992) 61, pp. 131–137.

59   Bercovitch, F. B., Berry, P. S. M.: 'Herd composition, kinship and fission-fusion social dynamics among wild giraffe', *African Journal of Ecology* (2013) 51, pp. 206–216.

60   Sundaresan, S. R., Fischhoff, I. R., Dushoff, J., Rubenstein, D. I.: 'Network metrics reveal differences in social organization between two fission-fusion species, Grevy's zebra and onager', *Oecologia* (2007) 151 (1), pp. 140–149.

61   Popa-Lisseanu, A. G., Bontadina, F., Mora, O., Ibáñez, C.: 'Highly structured fission-fusion societies in an aerial-hawking, carnivorous bat', *Animal Behaviour* (2008) 75 (2), pp. 471–482.

62   Croft, D. P., Krause, J., James, R.: 'Social networks in the guppy (*Poecilia reticulate*)', *Biology Letters* (2004) 271, pp. 516–519.

63   Lusseau, D. et al.: 'Quantifying the influence of sociality on population structure in bottlenose dolphins', *Journal of Animal Ecology* (2006) 75 (1), pp. 14–24.

64   Stanton, M. A., Gibson, Q. A., Mann, J.: 'When mum's away: a study of mother and calf ego networks during separations in wild bottlenose dolphins (*Tursiops* sp.)', *Animal Behaviour* (2011) 82, pp. 405–412.

65   Connor, R. C., Watson-Capps, J. J., Sherwin, W. B., Krützen, M.: 'A new level of complexity in the male alliance networks of Indian Ocean bottlenose dolphins (*Tursiops* sp.)', *Biology Letters* (2011) 7 (4), pp. 623–626.

66   Davidsen, J., Ebel, H., Bornholdt, S.: 'Emergence of a small world from

local interactions: modeling acquaintance networks', *Physical Review Letters* (2002) 88 (128701).

67 Barabasi, A. L., Albert, R.: 'Emergence of scaling in random networks', *Science* (1999) 286, pp. 509–512.

68 McComb, K., Moss, C., Sayailel, S., Baker, L.: 'Unusually extensive networks of vocal recognition in African elephants', *Animal Behavior* (2000) 59, pp. 1103–1109.

69 Kondo, N., Izawa, E. I., Watanabe, S.: 'Crows cross-modally recognize group members but not non-group members', *Proceedings of the Royal Society B* (2012) 279 (1735), pp. 1937–1942.

70 www.uni-erfurt.de/kit/.

71 McAuliffea, K., Jordan, J. J., Warneken, F.: 'Costly third-party punishment in young children', *Cognition* (2015) 134, pp. 1–10.

72 www.ted.com/talks/rebecca_saxe_how_brains_make_moral_judgments (from 3:50 onwards).

73 www.theguardian.com/world/2017/mar/15/recep-tayyip-erdogan-rails-against-dutch-in-televised-speech-netherlands-srebrenica.

74 www.youtube.com/watch?v=meiU6TxysCg.

75 Brosnan, S. F., de Waal, F. B. M.: 'Monkeys reject unequal pay', *Nature* (2003) 425, p. 297 ff.

76 Brosnan, S. F., Schiff, H. C., de Waal, F. B. M.: 'Tolerance for inequity may increase with social closeness in chimpanzees', *Proceedings of the Royal Society B* (2005) 272, pp. 253–258.

77 Brosnan, S. F., Flemming, T., Talbot, C. F., Mayo, L., Stoinski, T.: 'Orangutans (*Pongo pygmaeus*) do not form expectations based on their partner's outcomes', *Folia Primatologica* (2010) 82, pp. 56–70.

78 Talbot, C. F., Freeman, H. D., Williams, L. E., Brosnan, S. F.: 'Squirrel monkeys' response to inequitable outcomes indicates a behavioural convergence within the primates', *Biology Letters* (2011) 7, p. 680 ff.

79 Van Schaik, C. P., Damerius, L., Isler, K.: 'Wild orangutan males plan and communicate their travel direction one day in advance', *Public Library of Science One* (2013) 8 (9).

80 Krützen, M., Willems, E. P., van Schaik, C. P.: 'Culture and geographic variation in orangutan behaviour', *Current Biology* (2011) 21 (21).

81 Massen, J. J. M., van den Berg, L. M., Spruijt, B. M., Sterck, E. H. M.: 'Inequity aversion in relation to effort and relationship quality in long-tailed macaques (*Macaca fascicularis*)', *American Journal of Primatology* (2012) 74, pp. 145–156.

82 Range, F., Leitner, K., Viranyi, Z.: 'The influence of the relationship and

motivation on inequity aversion in dogs', *Social Justice Research* (2012) 25, pp. 170–194.

83   Wascher, C. A. F., Bugnyar, T.: 'Behavioral responses to inequity in reward distribution and working effort in crows and ravens', *Public Library of Science One* (2013) 8 (2), p. e56885. doi.org/10.1371/journal.pone.0056885.

84   Oberliessen, L. et al.: 'Inequity aversion in rats, *Rattus norvegicus*', *Animal Behaviour* (2016) 115, pp. 157–166.

85   Clay, Z., Ravaux, L., de Waal, F. B. M., Zuberbühler, K.: 'Bonobos (*Pan paniscus*) vocally protest against violations of social expectations', *Journal of Comparative Psychology* (2016) 130 (1), pp. 44–54.

86   Shaw, A., Olson, K. R.: 'Children discard a resource to avoid inequity', *Journal of Experimental Psychology: General* (2012) 141 (2), pp. 382–395.

87   Brosnan, S. F., Talbot, C., Ahlgren, M., Lambeth, S. P., Schapiro, S. J.: 'Mechanisms underlying responses to inequitable outcomes in chimpanzees, *Pan troglodytes*', *Animal Behaviour* (2010) 79 (6), pp. 1229–1237.

88   Leimgruber, K. L., Rosati, A. G., Santos, L. R.: 'Capuchin monkeys punish those who have more', *Evolution and Human Behavior* (2016) 37 (3), pp. 236–244.

89   Riedl, K., Jensen, K., Call, J., Tomasello, M.: 'No third-party punishment in chimpanzees', *Proceedings of the National Academy of Sciences* (2012) 109, pp. 14824–14829.

90   Haney, C., Banks, C., Zimbardo, P. G.: 'Interpersonal dynamics in a simulated prison', *International Journal of Criminology and Penology* (1973) 1, pp. 69–97.

91   Andrews, K: 'Understanding norms without a theory of mind', *Inquiry* (2009) 52 (5), pp. 433–448.

92   Bojanowski, E.: 'Vocal behaviour in bottlenose dolphins (*Tursiops truncatus*): ontogeny and contextual use in specific interactions.' Doctoral dissertation, Free University of Berlin, Germany 2002.

93   Gonçalves, A.: 'Blanket stealing in captive chimpanzees (*Pan troglodytes verus*): an observed case of spontaneous fairness related behavior', *Cadernos do GEEvH* (2015) 4 (1), pp. 25–40.

94   Thaler, R. H.: 'Toward a positive theory of consumer choice', *Journal of Economic Behavior and Organization* (1980) 1 (1), pp. 39–60.

95   Davies, N. B.: 'Territorial defence in the speckled wood butterfly (*Pararge aegeria*): the resident always wins', *Animal Behaviour* (1978) 26, pp. 138–147.

96   www.youtube.com/watch?v=CPznMbNcfO8, from 3:30 onwards.

97   Mitani, J. C., Watts, D. P., Amsler, S. J.: 'Lethal intergroup aggression leads

to territorial expansion in wild chimpanzees', *Current Biology* (2010) 20 (12), pp. R507–R508.

98   Goodall, J.: *The Chimpanzees of Gombe* (Cambridge, 1986).

99   Bradshaw, G. A. et al.: 'Elephant breakdown social trauma: early disruption of attachment can affect the physiology, behaviour and culture of animals and humans over generations', *Nature* (2005) 433.

100 McComb, K., Baker, L., Moss, C.: 'African elephants show high levels of interest in the skulls and ivory of their own species', *Biology Letters* (2006) 2, pp. 26 ff.

101 Douglas Hamilton, I., Bhalla, S., Wittemyer, G., Vollrath, F.. 'Behavioural reactions of elephants towards a dying and deceased matriarch', *Applied Animal Behaviour Science* (2006) 100 (1–2), pp. 87–102.

102 Ritesh, J: 'Social behaviour of Asian elephants. How social are Asian elephants *Elephas maximus*?', *New York Science Journal* (2010) 3 (1), pp. 27–31.

103 Goodall, J.: 'The behaviour of free-living chimpanzees in the Gombe Stream Reserve', *Animal Behaviour Monographs* (1968) 1, pp. 163–311.

104 Cronin, K. A., van Leeuwen, E. J. C., Mulenga, I. C., Bodamerm, M. D.: 'Behavioral response of a chimpanzee mother toward her dead infant', *American Journal of Primatology* (2011), 73 (5), pp. 415–421.

105 Kooriyama, T.: 'The death of a newborn chimpanzee at Mahale: reactions of its mother and other individuals to the body', *Pan Africa News* (2009) 16 (2).

106 Link to the article and various videos: www.sciencedirect.com/science/article/pii/S0960982210002186.

107 Biro, D. et al.: 'Chimpanzee mothers at Bossou, Guinea carry the mummified remains of their dead infants', *Current Biology* (2010) 20 (8), pp. R351–R352.

108 *'Nonetheless we hope that further data from this already threatened community will not be quick in coming.'*

109 Dudzinski, K. M., Sakai, M., Masaki, K., Kogi, K., Hishii, T., Kurimoto, M.: 'Behavioural observations of bottlenose dolphins towards two dead conspecifics', *Aquatic Mammals* (2003) 29 (1), pp. 108–116.

110 Ritter, F.: 'Behavioural responses of rough-toothed dolphins to a dead newborn calf', *Marine Mammal Science* (2007) 23 (2), pp. 429–433.

111 Kashkina, M. I.: '*Dendronasus* sp. a new member of the order nose-walkers (Rhinogradentia)', *Russian Journal of Marine Biology* (2004) 30 (2), pp. 148–150.

112 Stümpke, H.: *Bau und Leben der Rhinogradentia* (Stuttgart, 1961).

113 www.iucnredlist.org.

V. On thought

1   Proust, J.: 'Das intentionale Tier', in: Perler, D., Wild, M.: *Der Geist der Tiere* (Frankfurt/Main) 2005, pp. 223–244.

2   Troje, N. F., Huber, L., Loidolt, M., Aust, U., Fieder, M.: 'Categorical learning in pigeons: the role of texture and shape in complex static stimuli', *Vision Research* (1999) 39, pp. 353–366.

3   www.sciencedaily.com/releases/2009/02/090212141143.html.

4   Wu, W., Moreno, A. M., Tangen, J. M., Reinhard, J.: 'Honeybees can discriminate between Monet and Picasso paintings', *Journal of Comparative Physiology A* (2013) 199 (1), pp. 45–55.

5   Piaget, J.: 'The Development of Object Concept' in: Piaget, J.: *The Construction of Reality in the Child* (London 1999), pp. 3 – 96.

6   Pollok, B., Prior, H., Guentuerkuen, O.: 'Development of object permanence in food-storing magpies (*Pica pica*)', *Journal of Comparative Psychology* (2000) 114, pp. 148–157.

7   Miller, H. C., Gipson, C. D., Vaughan, A., Rayburn-Reeves, R., Zentall, T. R.: 'Object permanence in dogs: invisible displacement in a rotation task', *Psychonomic Bulletin & Review* (2009) 16, pp. 150–155.

8   Triana, E., Pasnak, R.: 'Object permanence in cats and dogs', *Animal Learning & Behavior* (1981) 9, pp. 135–139.

9   Gomez, J.: 'Species comparative studies and cognitive development', *Trends in Cognitive Sciences* (2005) 9, pp. 118–125.

10   Call, J.: 'Inferences about the location of food in the great apes (*Pan paniscus, Pan troglodytes, Gorilla gorilla*, and *Pongo pygmaeus*)', *Journal of Comparative Psychology* (2004) 118, pp. 232–241.

11   Schloegl, C., Schmidt, J., Boeckle, M., Weiß, B. M., Kotrschal, K.: 'Grey parrots use inferential reasoning based on acoustic cues alone', *Proceedings of the Royal Society B* (2012) 279 (1745).

12   O'Hara, M., Auersperg, A. M. I., Bugnyar, T., Huber, L.: 'Inference by exclusion in Goffin cockatoos (*Cacatua goffini*)', *Public Library of Science One* (2015) 10 (8).

13   O'Hara, M. et al.: 'Reasoning by exclusion in the kea (*Nestor notabilis*)', *Animal Cognition* (2016) 19, pp. 965.

14   Hill, A., Collier-Baker, E., Suddendorf, T.: 'Inferential reasoning by exclusion in children (*Homo sapiens*)', *Journal of Comparative Psychology* (2012) 126 (3), pp. 243–254.

15   Bräuer, J., Kaminski, J., Riedel, J., Call, J., Tomasello, M.: 'Making inferences about the location of hidden food: social dog, causal ape', *Journal of Comparative Psychology* (2006) 120, pp. 38–47.

16   Aust, U., Range, F., Steurer, M., Huber, L.: 'Inferential reasoning by exclusion in pigeons, dogs, and humans', *Animal Cognition* (2008) 11 (4), pp. 587–597.

17   Zaine, I., Domeniconi, C., de Rose, J. C.: 'Exclusion performance and learning by exclusion in dogs', *Journal of the Experimental Analysis of Behavior* (2016) 105 (3).

18   Taylor, A. H., Miller, R., Gray, R. D.: 'New Caledonian crows reason about hidden causal agents', *Proceedings of the National Academy of Sciences* (2012) 109 (40), pp. 16389–16391.

19   'Special Section: Reasoning versus association in animal cognition: current controversies and possible ways forward', *Journal of Comparative Psychology* (2016) 130 (3).

20   Herman, L. M., Richards, D. G., Wolz, J. P.: 'Comprehension of sentences by bottlenosed dolphins', *Cognition* (1984) 16, pp. 129–219.

21   Bates, L. A., Sayialel, K. N., Njiraini, N., Poole, J. H., Moss, C., Byrne, R. W.: 'African elephants have expectations about the locations of out-of-sight family members', *Biology Letters* (2008) 4, pp. 34 ff.

22   Martinho, I., Kacelnik, A: 'Ducklings imprint on the relational concept of »same or different', *Science* (2016) 353, p. 286.

23   www.cell.com/cms/attachment/2050817622/2059082563/mmc2.mp4.

24   Vonk, J.: Gorilla '(*Gorilla gorilla gorilla*) and orangutan (*Pongo abelii*) understanding of first- and second-order relations', *Animal Cognition* (2003) 6, pp. 77–86.

25   Flemming, T. M., Thompson, R. K. R., Fagot, J.: 'Baboons, like humans, solve analogy by categorical abstraction of relations', *Animal Cognition* (2013) 16, pp. 519–524.

26   Smirnova, A., Zorina, Z., Obozova, T., Wasserman, E.: 'Crows spontaneously exhibit analogical reasoning', *Current Biology* (2015) 25 (2), pp. 256–260.

27   Plutarch: *Plutarch's Morals* Translated from the Greek by several hands. Corrected and revised by Goodwin, W. W. (Boston, 1874), p. 163.

28   Cheke, L. G., Loissel, E., Clayton, N. S.: 'How do children solve Aesop's fable?', *Public Library of Science One* (2012) 7 (7), p. e40574. doi.org/10.1371/journal.pone.0040574.

29   Jelbert, S. A., Taylor, A. H., Cheke, L. G., Clayton, N. S., Gray, R. D.: 'Using the Aesop's fable paradigm to investigate causal understanding of water displacement by New Caledonian crows', *Public Library of Science One* (2014) 9 (3), p. e92895. doi.org/10.1371/journal.pone.0092895.

30   Ghirlandaa, S., Lindd, J.: '"Aesop's fable" experiments demonstrate trial-and-error learning in birds, but no causal understanding', *Animal Behaviour* (2017) 123, pp. 239–247.

31   Bird, C. D., Emery, N. J.: 'Rooks use stones to raise the water level to reach a floating worm', *Current Biology* (2009) 19, pp. 1410–1414.

32   Hanus, D., Mendes, N., Tennie, C., Call, J.: 'Comparing the performances of apes (*Gorilla gorilla*, *Pan troglodytes*, *Pongo pygmaeus*) and human children (*Homo sapiens*) in the floating peanut task', *Public Library of Science One* (2011) 6 (6), p. e19555. doi.org/10.1371/journal.pone.0019555.

33   Kuczaj, S. A., Gory, J. D., Xitco Jr., M. J.: 'How intelligent are dolphins? A partial answer based on their ability to plan their behavior when confronted with novel problems', *Japanese Journal of Animal Psychology* (2009) 59 (1), pp. 99–115.

34   Simila, T., Fugarte, F.: 'Surface and underwater observations of cooperatively feeding killer whales in northern Norway', *Canadian Journal of Zoology* (1993), 71, pp. 1494–1499.

35   Nottestad, L., Ferno, A., Axelsen, B. E.: 'Digging in the deep: killer whales' advanced hunting tactic', *Polar Biology* (2002) 25, pp. 939–941.

36   Guinet, C., Bouvier, J.: 'Development of intentional stranding hunting techniques in killer whale (*Orcinus orca*) calves at Crozet Archipelago', *Canadian Journal of Psychology* (1995) 73, pp. 27–33.

37   Visser, I. N., Smith, T. G., Bullock, I. D., Green, G. D., Carlsson, O. G., Imberti, S.: 'Antarctic peninsula killer whales (*Orcinus orca*) hunt seals and a penguin on floating ice', *Marine Mammal Science* (2008) 24, pp. 225–234.

38   Duffy-Echevarria, E. E., Connor, R. C., St. Aubin, D. J.: 'Observations of strandfeeding behavior by bottlenose dolphins (*Tursiops truncatus*) in Bull Creek, South Carolina', *Marine Mammal Science* (2008) 24 (1), pp. 202–206.

39   Fertl, D., Wilson, B.: 'Bubble use during prey capture by a lone bottlenose dolphin (*Tursiops truncatus*)', *Aquatic Mammals* (1997) 23 (2), p. 113 f.

40   Lewis, J. S., Schroeder, W.: 'Mud plume feeding, a unique foraging behavior of the bottlenose dolphin (*Tursiops truncatus*) in the Florida Keys, Gulf of Mexico', *Science* (2003) 21 (1).

41   Smolker, R. A., Richards, A. F., Connor, R. C., Mann, J., Berggren, P.: 'Sponge carrying by Indian Ocean bottlenose dolphins: possible tool use by a delphinid', *Ethology* (1997) 103, pp. 454–465.

42   Pryor, K., Lindbergh, J., Lindbergh, S., Milano, R.: 'A dolphin-human fishing cooperative in Brazil', *Marine Mammal Science* (1990) 6, pp. 77– 82.

43   Onishi, S.: 'Mutualistic fishing between fisherman and Irrawaddy dolphins in Myanmar', *Tigerpaper* (2008) 35, pp. 1–8.

44   Guillerault, N. et al.: 'Does the non-native European catfish *Silurus glanis* threaten French river fish populations?', *Freshwater Biology* (2015) 60 (5), pp. 922–928.

45 The discipline of epigenetics concerns the environmentally conditioned activation and deactivation of genes, processes in which the genetic code is not altered.

46 Pruetz, J. D., Bertolani, P.: 'Savanna chimpanzees, *Pan troglodytes verus*, hunt with tools', *Current Biology* (2007) 17 (5), pp. 412–417.

47 Boesch, C.: 'Joint cooperative hunting among wild chimpanzees: taking natural observations seriously. Commentary/Tomasello et al.: Understanding and sharing intentions', *Behavioral and Brain Sciences* (2005) 28 (5).

48 en.wikipedia.org/wiki/Number_sense_in_animals.

49 www.guardian.co.uk/science/2003/jul/03/research.science/print.

50 www.youtube.com/watch?v=Y7kjsb7iyms.

51 Sellitto, M., Ciaramelli, E., di Pellegrino, G.: 'The neurobiology of intertemporal choice: insight from imaging and lesion studies', *Reviews in the Neurosciences* (2011) 22 (5).

52 Mischel, W.: *The Marshmallow Test: Understanding Self-Control and How to Master It* (London, 2015).

53 Call, J., Carpenter, M.: 'Do apes and children know what they have seen?', *Animal Cognition* (2001) 3 (4), pp. 207–220.

54 Foote, A. L., Crystal, J. D.: 'Metacognition in the rat', *Current Biology* (2007) 17 (6), pp. 551–555.

55 Haun, D. B. M., Nawroth, C., Call, J.: 'Great apes' risk-taking strategies in a decision making task', *Public Library of Science One* (2011) 6 (12), p. e28801. Doi:10.1371/journal.pone.0028801.

56 Smith, J. D., Schull, J., Strote, J., McGee, K., Egnor, R., Erb, L.: 'The uncertain response in the bottlenosed dolphin (*Tursiops truncatus*)', *Journal of Experimental Psychology: General* (1995) 124 (4), pp. 391–408.

57 Rosati, A. G., Santos, L. R.: 'Spontaneous metacognition in rhesus monkeys', *Psychological Science* (2016) 27 (9).

58 Vining, A. Q., Marsh, H. L.: 'Information seeking in capuchins (*Cebus apella*): a rudimentary form of metacognition?', *Animal Cognition* (2015) 18 (3), pp. 667–681.

59 Castro, L., Wasserman, E. A.: 'Information-seeking behavior: exploring metacognitive control in pigeons', *Animal Cognition* (2013) 16, pp. 241–254.

60 Perry, C. J., Barron, A. B.: 'Honey bees selectively avoid difficult choices', *Proceedings of the National Academy of Sciences of the United States of America* (2013) 110 (47), pp. 19155–19159.

61 Broom, D. M., Sena, H., Moynihan, K. L.: 'Pigs learn what a mirror image represents and use it to obtain information', *Animal Behaviour* (2009) 78 (5), pp. 1037–1041.

62   Itakura, S.: 'Mirror guided behavior in Japanese monkeys (*Macaca fuscata fuscata*)', *Primates* (1987) 28, pp. 149–161.

63   Pepperberg, I. M., Garcia, S. E., Jackson, E. C., Marconi, S.: 'Mirror use by African grey parrots (*Psittacus erithacus*)', *Journal of Comparative Psychology* (1995) 109, pp. 182–195.

64   Medina, F. S., Taylor, A. H., Hunt, G. R., Gray, R. D.: 'New Caledonian crows' responses to mirrors', *Animal Behaviour* (2011) 82, pp. 981–993.

65   Howell, T. J., Bennett, P. C.: 'Can dogs (*Canis familiaris*) use a mirror to solve a problem?', *Journal of Veterinary Behavior: Clinical Applications and Research* (2011) 6 (6), pp. 306–312.

66   Parker, S. T.: 'A developmental approach to the origins of self-recognition in great apes and human infants', *Journal of Human Evolution* (1991) 6, pp. 435–449.

67   Gallup Jr., G. G.: 'Chimpanzees: self recognition', *Science* (1970) 167 (3914), p. 86 f.

68   Patterson, F. G., Cohn, R. H.: 'Self-awareness.' In: Parker, S. T., Mitchell, R. W., Boccia, M. L.: *Animals and Humans. Developmental Perspectives* (Cambridge, 1994), pp. 273–290.

69   Hyatt, C. W.: 'Responses of gibbons (*Hylobates lar*) to their mirror images', *American Journal of Primatology* (1998) 45, pp. 30–311/Anderson, J. R.: 'Responses to mirror image stimulation and assessment of self-recognition in miror- and peer-reared stumptail macaques', *Quarterly Journal of Experimental Psychology* (1983) 35 (3), pp. 201–212/Bayart, F., Anderson, J. R.: 'Mirror-image reactions in a tool-using, adult male *Macaca tonkeana*', *Behavioural Processes* (1985) 10 (3), pp. 219–227/Gallup, G. G., Wallnau, L., Suarez, S. D.: 'Failure to find self-recognition in mother-infant and infant-infant rhesus monkey pairs', *Folia Primatologica* (1980) 33 (3), pp. 210–219/Suarez, S. D., Gallup, G. G. Jr.: 'Social responding to mirrors in rhesus macaques: effects of changing mirror location', *American Journal of Primatology* (1986) 11, pp. 239–244/Anderson, J. R., Roeder, J. J.: 'Responses of capuchin monkeys (*Cebus apella*) to different conditions of mirror-image stimulation', *Primates* (1989) 30 (4), pp. 581–587/Povinelli, D. J.: 'Failure to find self-recognition in Asian elephants (*Elephas maximus*) in contrast to their use of mirror cues to discover hidden food', *Journal of Comparative Psychology* (1989) 103 (2), pp. 122–131.

70   Reiss, D., Marino, L.: 'Mirror self-recognition in the bottlenose dolphin: a case of cognitive convergence', *Proceedings of the National Academy of Sciences* (2001) 98 (10), pp. 5937–5942.

71 Delfoura, F., Marten, K.: 'Mirror image processing in three marine mammal species: killer whales (*Orcinus orca*), false killer whales (*Pseudorca crassidens*) and California sea lions (*Zalophus californianus*)', *Behavioural Processes* (2001) 53 (3), pp. 181–190.

72 Plotnik, J. M. P., de Waal, F. B. M., Reiss, D.: 'Self-recognition in an Asian elephant', *Proceedings of the National Academy of Sciences* (2006) 103 (45), pp. 17053–17057.

73 Prior, H., Schwarz, A., Güntürkün, O., de Waal, F. B. M.: 'Mirror-induced behavior in the magpie (*Pica pica*): evidence of self-recognition', *PLoS Biology* (2008) 6 (8).

74 Rahde, T.: 'Stufen der mentalen Repräsentation bei Keas (*Nestor notabilis*).' Dissertation im Fachbereich Biologie, Chemie, Pharmazie der Freien Universität Berlin (2014) www.diss.fu-berlin.de/diss/receive/FUDISS_thesis_000000096348.

75 www.youtube.com/watch?v=M2IokwSua44.

76 www.wired.com/wiredscience/2010/09/monkey-self-awareness/.

77 Rajala, A. Z., Reininger, K. R., Lancaster, K. M., Populin, L. C.: 'Rhesus monkeys (*Macaca mulatta*) do recognize themselves in the mirror: implications for the evolution of self-recognition', *Public Library of Science One* (2010) 5 (9), p. e12865. doi.org/10.1371/journal.pone.0012865.

78 Epstein, L., Skinner, R. P., Skinner, B. F.: 'Self-awareness in the pigeon', *Science* (1981) 212 (4495), p. 695 f.

79 Lewis, M.: 'The origins and uses of self awareness or the mental representation of me', *Consciousness and Cognition* (2011) 20, pp. 120–129.

80 Broesch, T., Callaghan, T., Henrich, J., Murphy, C., Rochat, P.: 'Cultural variations in children's mirror self-recognition', *Journal of Cross-Cultural Psychology* (2011) 42 (6), pp. 1018–1029.

81 Asendorpf, J. B., Warkentin, V., Baudonniere, P. M.: 'Self-awareness and other awareness II: mirror self-recognition, social contigency awareness, and synchronic imitation', *Developmental Psychology* (1996) 32 (2), pp. 313–321.

82 Derégnaucourt, S., Bovet, D.: 'The perception of self in birds', *Neuroscience and Biobehavioral Reviews* (2016) 69, pp. 1–14.

83 Ari, C., D'Agostino, D. P.: 'Contingency checking and self-directed behaviors in giant manta rays: do elasmobranchs have self-awareness?', *Journal of Ethology* (2016) 34 (2), pp. 167–174.

84 Cammaerts, M. C., Cammaerts, R.: 'Are ants (Hymenoptera, Formicidae) capable of self recognition?', *Journal of Science* (2015) 5, pp. 521–532.

85 Schetsche, M.: *Interspezies-Kommunikation. Voraussetzungen und Grenzen.*

PeriLog – Freiburger Beiträge zur Kultur- und Sozialforschung (Berlin, 2014).

86   Hodson, H.: 'I know it's me talking', *New Scientist* (2015) 18.

87   Caldwell, M. C., Caldwell, D. K.: 'Individualized whistle contours in bottle-nose dolphins (*Tursiops truncatus*)', *Science* (1965) 207, p. 434 f. See also Caldwell, M. C., Caldwell, D. K., Tyack, P. L.: 'Review of the signature-whistle hypothesis for the Atlantic bottlenose dolphin.' In: Leatherwood, S., Reeves, R. R.: *The Bottlenose Dolphin* (New York, 1990), pp. 199–233.

88   Quick, N. J., Janik, V. M.: 'Bottlenose dolphins exchange signature whistles when meeting at sea', *Proceedings of the Royal Society* (2012) 279 (1738), pp. 2539–2545.

89   In the so-called SOFAR (SOund Fixing And Ranging) channel, which is located at a depth of 500 to 1,000 metres, depending on temperature, pressure and the salt content of the water, sound has particularly low resistance and can be conducted extremely well. Sperm whales are known to use this channel to navigate across several thousands of kilometres of the ocean.

90   King, S. L., Janik, V.: 'Bottlenose dolphins can use learned vocal labels to address each other', *Proceedings of the National Academy of Sciences of the United States of America* (2013) 110 (32), pp. 13216–13221.

91   Watwood, S. L., Owen, E. C. G., Tyack, P. L., Wells, R. S.: 'Signature whistle use by temporarily restrained and free-swimming bottlenose dolphins, *Tursiops truncatus*', *Animal Behaviour* (2005) 69, pp. 1373–1386.

92   Richards, D. G., Wolz, J. P., Herman, L. M.: 'Vocal mimicry of computer-generated sounds and vocal labeling of objects by a bottlenosed dolphin, *Tursiops truncatus*', *Journal of Comparative Psychology* (1984) 98, pp. 10–28.

93   Herman, L. M., Richards, D. G., Wolz, J. P.: 'Comprehension of sentences by bottlenosed dolphins', *Cognition* (1984) 16, pp. 129–219.

94   Janik, V.: 'Cetacean vocal learning and communication', *Current Opinion in Neurobiology* (2014) 28, pp. 60–65.

95   Watwood, S. L.; Tyack, P. L.; Wells, R. S.: 'Whistle sharing in paired male bottlenose dolphins, *Tursiops truncatus*', *Behavioral Ecology and Sociobiology* (2004) 55, pp. 531–543.

96   Berg, K. S., Delgado, S., Okawa, R., Beissinger, S. R., Bradbury, J. W.: 'Contact calls are used for individual mate recognition in free-ranging green-rumped parrotlets, *Forpus passerinus*', *Animal Behaviour* (2011) 81, pp. 241–248.

97   Janik, V., Sayigh L. S.: 'Communication in bottlenose dolphins: 50 years of signature whistle research', *Journal of Comparative Physiology A* (2013) 199, pp. 479–489.

98 Berg, K. S., Delgado, S., Cortopassi, K. A., Beissinger, S. R., Bradbury, J. W.: 'Vertical transmission of learned signatures in a wild parrot', *Proceedings of the Royal Society B* (2012) 279, pp. 585–591.

99 Brensing, K: *Persönlichkeitsrechte für Tiere: Die nächste Stufe der moralischen Evolution* (Freiburg, 2013).

100 Umami is often described as 'savoury'. Our senses of taste react primarily to the amino acid glutamine, which is present as a 'substitute protein' in all protein-rich foods.

101 John O. P., Naumann, L. P., Soto, C. J.: 'Paradigm Shift to the Integrative Big Five Trait Taxonomy.' In: *Handbook of Personality Theory and Research*, Third Edition (2008).

102 Dingemanse, N. J., Réale, D.: 'Natural selection and animal personality', *Behaviour* (2005) 142 (9–10), pp. 1159–1168.

103 Holbrook, C. T., Wright, C. M., Pruitt, J. N.: 'Individual differences in personality and behavioural plasticity facilitate division of labour in social spider colonies', *Animal Behaviour* (2014) 97, pp. 177–183.

104 Briffa, M., Sneddon, L. U.: 'Proximate mechanisms of animal personality among-individual behavioural variation in animals', *Behaviour* (2016) 153 (13–14), pp. 1509–1515.

105 Jones, A. C., Gosling, S. D.: 'Temperament and personality in dogs (*Canis familiaris*): a review and evaluation of past research', *Applied Animal Behaviour Science* (2005) 95 (1), pp. 1 53.

106 Briffa, M., Weiss, A.: 'Animal personality', *Current Biology* (2010) 20, pp. R912–R914.

107 Kandler, C., Riemann, R., Spinath, F. M., Angleitner, A.: 'Sources of variance in personality facets: a multiple-rater twin study of self–peer, peer–peer, and self self (dis)agreement', *Journal of Personality* (2010) 78 (5), pp. 1565–1594.

108 Verhulst, C. E., Mateman, A. C., Zwier, M. V., Caro, S. P., Verhoeven, K. J. F., van Oers, K.: 'Evidence from pyrosequencing indicates that natural variation in animal personality is associated with DRD4 DNA methylation', *Molecular Ecology* (2016) 25, pp. 1801–1811.

109 Brennecke, A. et al.: *Biosphäre Sekundarstufe II – Themenbände: Ökologie. Schülerbuch* (Berlin, 2012).

110 www2.klett.de/sixcms/media.php/82/biomax_epigenetik.pdf.

111 Bartal, I. B. A., Decety, J., Mason, P.: 'Empathy and pro-social behavior in rats', *Science* (2011) 334 (6061), pp. 1427–1430.

112 Heinzen, T. E., Lilienfeld, S. O., Nolan, S. A.: 'Clever Hans. What a horse can teach us about self-deception', *Skeptic* (2015) 20 (1), pp. 10–18.

113   Ibid.

114   Krall, K.: *Denkende Tiere. Beiträge zur Tierseelenkunde auf Grund eigener Versuche. Der kluge Hans und meine Pferde Muhamed und Zarif* (Leipzig, 1912).

115   de.wikipedia.org/wiki/Carl_Georg_Schillings.

116   Malavasi, R., Huber, L.: 'Evidence of heterospecific referential communication from domestic horses (*Equus caballus*) to humans', *Animal Cognition* (2016) 19 (5).

117   Premack, D., Woodruff, G.: 'Does the chimpanzee have a theory of mind?', *Behavioral and Brain Sciences* (1978) 1, S: 515–526.

118   www.eco-etho-recherche.com.

119   Kiley-Worthington, M.: 'Nonhuman mind-reading ability. Commentary on Harnad on Other Minds', *Animal Sentience* (2016) 2016.070.

120   Chen, Q., Panksepp, J. B., Lahvis, G. P.: 'Empathy is moderated by genetic background in mice', *Public Library of Science One* (2009) 4 (2).

121   Dallya, J. M., Emery, N. J., Clayton, N. S.: 'Avian theory of mind and counter espionage by food-caching western scrub-jays (*Aphelocoma californica*)', in Special Issue: Theory of mind: specialized capacity or emergent property? *European Journal of Developmental Psychology* (2010) 7 (1), pp. 17–37.

122   saxelab.mit.edu/index.php.

123   Call, J., Tomasello, M.: 'Does the chimpanzee have a theory of mind? 30 years later', *Trends in Cognitive Sciences* (2008) 12 (5).

124   www.youtube.com/watch?v=dawfSPx3yPM See also: www.you-tube.com/watch?v= Mol29ghH2GE.

125   Krupenye, C., Kano, F., Hirata, S., Call, J., Tomasello, M.: 'Great apes anticipate that other individuals will act according to false beliefs', *Science* (2016) 354 (6308), pp. 110–114.

126   Southgate, V., Senju, A., Csibra, G.: 'Action anticipation through attribution of false belief by 2-year-olds', *Psychological Science* (2007) 18, pp. 587–592

127   www.youtube.com/watch?feature=player_embedded&v=CCXx2bNk6UA.

128   www.youtube.com/watch?v=h7XjOMUpa1I.

129   Caldwell, M. C., Caldwell, D. K.: 'Epimeletic (care-giving) behavior in cetacea', Chapter 33, S: 755–788. In: Norris, K. S. (ed.). *Whales, Dolphins and Porpoises* (Berkeley, 1966).

130   Pilleri, G.: 'Epimeletic behavior in cetacea: Intelligent or instinctive?' In: Pilleri, G.: *Investigations on Cetacea* (Sastamala, 1984), pp. 30–48.

131   www.youtube.com/watch?v=-lw8_SAtX80.

132 Connor, R. C., Norris, K. S.: 'Are dolphins reciprocal altruists?', *The American Naturalist* (1982) 119 (3), pp. 372–385.

133 Bates, L. A. et al.: 'Do elephants show empathy?', *Journal of Consciousness Studies* (2008) 15 (10–11), pp. 204–225.

134 Kahneman, D., Tversky, A.: 'Prospect theory: an analysis of decision under risk', *Econometrica* (1979) 47, pp. 263–292.

135 Chen, M. K., Lakshminarayanan, V., Santos, L. R.: 'The evolution of our preferences: evidence from capuchin monkey trading behavior', *Journal of Political Economy* (2006) 114, pp. 517–537.

136 Latty, T., Beekman, M.: 'Irrational decision-making in an amoeboid organism: transitivity and context-dependent preferences', *Proceedings of the Royal Society B* (2010) 278 (1703).

137 Reida, C. R., Lattya, T., Dussutourc, A., Beekmana, M.: 'Slime mold uses an externalized spatial "memory" to navigate in complex environments', *Proceedings of the National Academy of Sciences* (2012) 109 (43), pp. 17490–17494.

138 *'If I am saving you from turning the wrong way when you are lost, what difference does it make if you don't know I am steering you?'* www.mondora.com/#!/post/410b0d1b157ffa2dd33f43e504841b66.

139 www.ted.com/talks/lang/en/vilayanur_ramachandran_on_your_mind.html.

140 Gazzaniga, M. S: *Who's in Charge?: Free Will and the Science of the Brain* (New York, 2011).

141 Soon, C. S., Brass, M., Heinze, H. J., Haynes, J. D.: 'Unconscious determinants of free decisions in the human brain', *Nature Neuroscience* (2008) 11 (5), p. 543 ff.

142 Soon, C. S., He, A. H., Bode, S., Haynes, J. D.: 'Predicting free choices for abstract intentions', *Proceedings of the National Academy of Sciences* (2013) 110 (15).

143 Romero, T., Ito, M., Saito, A., Hasegawa, T.: 'Social modulation of contagious yawning in wolves', *Public Library of Science One* (2014) 9 (8) p. E105963. doi.org/10.1371/journal.pone.0105963.

144 Miller, M. L., Gallup, A. C., Vogel, A. R., Vicario, S. M., Clark, A. B.: 'Evidence for contagious behaviors in budgerigars (*Melopsittacus undulatus*): an observational study of yawning and stretching', *Behavioural Processes* (2012) 89, pp. 264–270.

145 Joly-Mascheroni, R. M., Senju, A., Shepherd, A. J.: 'Dogs catch human yawns', *Biology Letters* (2008) 4, p. 446 ff.

146 Keysers, C., Gazzola, V.: 'A plea for cross-species social neuroscience', *Current Topics in Behavioral Neurosciences* (2017) 30, pp. 179–191.

147   Lamma, C., Majdandžića, J.: 'The role of shared neural activations, mirror neurons, and morality in empathy – a critical comment', *Neuroscience Research* (2015) 90, pp. 15–24.

148   Allman, J. M., Watson, K. K., Tetrault, N. A., Hakeem, A. Y.: 'Intuition and autism: a possible role for von Economo neurons', *Trends in Cognitive Science* (2005) 9, pp. 367–373.

149   Nimchinsky, E. A., Gilissen, E., Allman, J. M., Perl, D. P., Erwin, J. M., Hof, P. R.: 'A neuronal morphologic type unique to humans and great apes', *Proceedings of the National Academy of Sciences* (1999) 96, pp. 5268–5273.

150   Butti, C., Sherwood, C. C., Hakeem, A. Y., Allman, J. M., Hof, P. R.: 'Total number and volume of von Economo neurons in the cerebral cortex of cetaceans', *Journal of Comparative Neurology* (2009) 515 (2), pp. 243–259.

151   Hakeem, A. Y., Sherwood, C. C., Bonar, C. J., Butti, C., Hof, P. R., Allman, J. M.: 'Von Economo neurons in the elephant brain', *Anatomical Record: Advances in Integrative Anatomy and Evolutionary Biology* (2008) 292 (2), pp. 242–248.

152   Güntürkün, O., Bugnyar, T.: 'Cognition without Cortex', *Trends in Cognitive Sciences* (2016) 20 (4), pp. 291–303.

153   Prathera, J., Okanoyab, K., Bolhuis, J. J.: 'Brains for birds and babies: neural parallels between birdsong and speech acquisition', *Neuroscience & Biobehavioral Reviews* (2017) accepted manuscript in press. doi.org/10.1016/j. neubiorev.2016.12.035.

154   Olson, C. R., Owen, D. C., Ryabinin, A. E., Mello, C. V.: 'Drinking songs: alcohol effects on learned song of zebra finches', *Public Library of Science One* (2014). dx.doi.org/10.1371/journal.pone.0115427.

155   Bshary, R., Gingins, S., Vail, A. L.: 'Social cognition in fishes', *Trends in Cognitive Sciences* (2014) 18 (9), pp. 465–471.

156   www.welt.de/vermischtes/article114579619/Kamtschatkas-Drogen-Baeren-sind-kerosinsuechtig.html.

157   Nikolaenko, V. A.: *Kamchatka Bear* (Moscow, 2003).

158   Seryodkin, I. V.: 'Marking activity of the Kamchatka brown bear (Ursus arctos piscator)', *Achievements in the Life Sciences* (2014) 8 (2), pp. 153–161.

159   Shohat-Ophir, G., Kaun, K. R., Azanchi, R., Mohammed, H., Heberlein, U.: 'Sexual deprivation increases ethanol intake in Drosophila', *Science* (2012) 335, pp. 1351–1355.

160   Kuo, L. E. et al.: 'Neuropeptide Y acts directly in the periphery on fat tissue and mediates stress-induced obesity and metabolic syndrome', *Nature Medicine* (2007) 13, pp.803–811.

161   Thiele, T. E., Koh, M. T., Pedrazzini, T.: 'Voluntary alcohol consumption is controlled via the neuropeptide Y Y1 receptor', *Journal of Neuroscience* (2002) 22 (3).

162   Guevara-Fiore, P., Endler, J. A.: 'Male sexual behaviour and ethanol consumption from an evolutionary perspective: a comment on "sexual deprivation increases ethanol intake in Drosophila"', *Fly* (2014) 8 (4), p. 234 ff.

163   www.youtube.com/watch?v=5otlF3kGbT4.

164   drinksint.com/news/fullstory.php/aid/4679/Amarula_Trust_funds_elephant_protection_project_. html.

165   Morris, S., Humphreys, D., Reynolds, D.: Myth, Marula, and elephant: 'an assessment of voluntary ethanol intoxication of the african elephant (*Loxodonta africana*) following feeding on the fruit of the marula tree (*Sclerocarya birrea*)', *Physiological and Biochemical Zoology* (2016) 79 (2).

166   news.bbc.co.uk/2/hi/south_asia/3423881.stm.

167   news.bbc.co.uk/2/hi/south_asia/2583891.stm.

168   news.bbc.co.uk/2/hi/asia-pacific/8118257.stm.

169   Pfister, J. A., Stegelmeier, B. L., Gardner, D. R., James, L. F.: 'Grazing of spotted locoweed (*Astragalus lentiginosus*) by cattle and horses in Arizona', *Journal of Animal Science* (2003) 81 (9), pp. 2285–2293.

170   Dudley, R.: 'Fermenting fruit and the historical ecology of ethanol ingestion: is alcoholism in modern humans an evolutionary hangover?', *Addiction* (2002) 97, pp. 381–388.

171   Heil, M. et al.: 'Partner manipulation stabilises a horizontally transmitted mutualism', *Ecology* Letters (2014) 17 (2) pp. 185–192.

172   Sueda, K. L. C., Hart, B. L., Cliff, K. D.: 'Characterisation of plant eating in dogs', *Applied Animal Behaviour Science* (2008) 111, pp. 120–132.

173   Strompfová, V. et al.: 'Experimental application of *Lactobacillus fermentum* CCM 7421 in combination with chlorophyllin in dogs', *Applied Microbiology and Biotechnology* (2015) 99, pp. 8681–8690.

174   Laurimaa, L. et al.: 'Alien species and their zoonotic parasites in native and introduced ranges: the raccoon dog example', *Veterinary Parasitology* (2016) 219, pp. 24–33.

175   The term 'bag' in this context denotes the total number of game animals killed within a particular hunting ground over a specified period.

176   www.jagdverband.de/sites/default/files/2015_Jahresjagdstrecke%20Marderhund_13_14.pdf.

177   Bos, N., Sundstrom, L., Fuchs, S., Freitak, D.: 'Ants medicate to fight disease', *Evolution* (2015) 69 (11), pp. 2979–2984.

178   Gilardi, J. D. et al.: 'Biochemical functions of geophagy in parrots: detoxi-

fication of dietary toxins and cytoprotective effects', *Journal of Chemical Ecology* (1999) 25 (4), pp. 897–922.

179 Revis, H. C., Waller, D. A.: 'Bactericidal and fungicidal activity of ant chemicals on feather parasites: an evaluation of anting behavior as a method of selfmedication in songbirds', *The Auk* (2004) 121 (4), pp. 1262–1268.

180 Birkinshaw, C. R.: 'Use of millipedes by black lemurs to anoint their bodies', *Folia Primatologica* (1999) 70, p. 170 f.

181 www.youtube.com/watch?v=iJoYlRH1Xdo.

182 Shuker, K. P. N.: *The Hidden Powers of Animals: Uncovering the Secrets of Nature* (London, 2001).

183 Shurkin, J.: 'News feature: Animals that self-medicate', *Proceedings of the National Academy of Sciences* (2014) 111 (49), pp. 17339–17341.

184 Suárez-Rodríguez, M., López-Rull, I., Garcia, C. M.: 'Incorporation of cigarette butts into nests reduces nest ectoparasite load in urban birds: new ingredients for an old recipe?', *Biology Letters* (2013) 9 (1).

185 Hart, B. L.: 'Behavioral adaptations to pathogens and parasites: five strategies', *Neuroscience & Biobehavioral Reviews* (1990) 14, pp. 273–294.

186 Freeland, W. J.: 'Pathogens and the evolution of primate sociality', *Biotropica* (1976) 8, pp. 12–24.

187 Hart, B. L.: 'Behavioral defenses in animals against pathogens and parasites: parallels with the pillars of medicine in humans', *Philosophical Transaction of the Royal Society B* (2011) 366, pp. 3406–3417.

VI: Sentimentality

1   Olds, J., Milner, P.: 'Positive reinforcement produced by electrical stimulation of septal area and other regions of rat brain', *Journal of Comparative and Physiological Psychology* (1954) 47 (6), pp. 419–427.

2   Berridge, K. C.: 'Food reward: brain substrates of wanting and liking', *Neuroscience & Biobehavioral Reviews* (1996) 20 (1), pp. 1–25.

3   Tye, K. M. et al.: 'Dopamine neurons modulate neural encoding and expression of depression-related behavior', *Nature* (2013) 493, pp. 537–541.

4   Ferris, C. F., Kulkarni, P., Sullivan, J. M., Harder, J. A., Messenger, T. L., Febo, M.: 'Pup suckling is more rewarding than cocaine: evidence from functional magnetic resonance imaging and three-dimensional computational analysis', *Journal of Neuroscience* (2005) 25, pp. 149–156.

5   www.ons.gov.uk/peoplepopulationandcommunity/healthandsocialcare/ causesofdeath/bulletins/alcoholrelateddeathsintheunitedkingdom/ registeredin2017.

6   Key, B.: 'Why fish do not feel pain', *Animal Sentience* (2016) 2016.003.

7   www.smithsonianmag.com/science-nature/fish-feel-pain-180967764/.

8   de.statista.com/statistik/daten/studie/153178/umfrage/konsum-von-antidepressiva-in-ausgwaehlten-laendern/.

9   'Beeinflussung des Kampfverhaltens von *Betta splendens* durch Psycho-pharmaka', *Advances in Ethology* (1984) 66 (S26), pp. 42–77 DOI: 10.1111/j.1439-0310.1984.tb00238.x.

10   Soares, M. C., Oliveira, R. F., Ros, A. F., Grutter, A. S., Bshary, R.: 'Tactile stimulation lowers stress in fish', *Nature Communications* (2011) 2, p. 534.

11   Brandl, S. J., Bellwood, D. R.: 'Coordinated vigilance provides evidence for direct reciprocity in coral reef fishes', *Scientific Reports* (2015) 5 (14556).

12   Burgdorf, P. J.: '"Laughing" rats and the evolutionary antecedents of human joy?', *Physiology & Behavior* (2005) 79 (3), pp. 533–547.

13   Simonet, P., Versteeg, D., Storie, D.: 'Dog-laughter: recorded playback reduces stress related behavior in shelter dogs', Proceedings of the 7th International Conference on Environmental Enrichment July 31–August 5 (2005).

14   Panksepp, J.: 'Beyond a joke: from animal laughter to human joy?', *Science* (2005) 308 (5718), p. 62 f.

15   Bowlby, J.: *Child care and the growth of love* (London, 1953).

16   Hernandez-Lallement, J., van Wingerden, M., Kalenscher, T.: 'Towards an animal model of callousness', *Neuroscience and Biobehavioral Reviews* (2016) pii: S0149-7634(16)30124-5.

VII: The pinnacle of creation

1   Gould, T. D. et al.: 'Animal models to improve our understanding and treatment of suicidal behavior', *Translational Psychiatry* (2017) 7, p. e1092.

2   Ching, E: 'The complexity of suicide: review of recent neuroscientific evidence', *Journal of Cognition and Neuroethics* (2016) 3 (4), pp. 27–40.

3   eu.usatoday.com/story/news/world/2017/07/03/g-20-germany-angela-merkel-russia-vladimir-putin/438376001/.

4   Haun, D. B. M., Rekers, Y., Tomasello, M.: 'Children conform to the behavior of peers; other great apes stick with what they know', *Psychological Science* (2014) 25 (12), pp. 2160–2167.

5   www.ted.com/talks/yuval_noah_harari_what_explains_the_rise_of_humans#t-461325.

6   Herculano-Houzel, S.: 'The human brain in numbers: a linearly scaled-up primate brain', *Frontiers in Human Neuroscience* (2009). doi.org/10.3389/neuro.09.031.2009.

7   Humphrey, N. K.: 'The social function of intellect.' In: Bateson, Hinde (eds.) *Growing Points in Ethology* (Cambridge University Press, 1976), pp. 303–317.

8   Seyfarth, R. M., Cheney, D. L.: 'What are big brains for?', *Proceedings of the National Academy of Sciences of the United States of America* (2002) 99, p. 4141 f.

9   Benson-Amram, S. et al.: 'Brain size predicts problem-solving ability in mammalian carnivores', *Proceedings of the National Academy of Sciences of the United States of America* (2016) 113 (9), pp. 2532–2537.

10   Evan, L., MacLean, E. L. et al.: 'The evolution of self-control', *Proceedings of the National Academy of Sciences of the United States of America* (2014) 111 (20), pp. E2140–E2148.

11   Miller, J. A., Horvath, S., Geschwind, D. H.: 'Divergence of human and mouse brain transcriptome highlights Alzheimer disease pathways', *Proceedings of the National Academy of Sciences of the United States of America* (2010) 107 (28), pp. 12698–12703.

12   Gácsi, M. et al.: 'Humans attribute emotions to a robot that shows simple behavioural patterns borrowed from dog behavior', *Computers in Human Behavior* (2016) 59, pp. 411–419.

13   Turing, A. M.: Computing machinery and intelligence. *Mind* (1950) 59 (236), pp. 433–460.

14   www.youtube.com/watch?v=VTNmLt7QX8E.

15   Heider, F., Simmel, M.: 'An experimental study of apparent behavior', *American Journal of Psychology* (1944) 57 (2), p. 243.

16   Dymond, S., Stewart, I.: 'Relational and analogical reasoning in comparative cognition', *International Journal of Comparative Psychology* (2016) 29.

17   Harley, H. E.: 'Consciousness in dolphins? A review of recent evidence', *Journal of Comparative Physiology A* (2013) 199 (6), pp. 565–582.

18   Leavens, D. A. et al.: 'Distal communication by chimpanzees (*Pan troglodytes*): evidence for common ground?', *Child Development* (2015) 86 (5), pp. 1623–1638.

19   Boesch, C.: 'Joint cooperative hunting among wild chimpanzees: taking natural observations seriously. Commentary/Tomasello et al.: Understanding and sharing intentions', *Behavioral and Brain Sciences* (2005) 18.

20   Eisfeld, S. M., Simmonds, M. S., Stansfield, L. R.: 'Behavior of a solitary sociable female bottlenose dolphin (*Tursiops truncatus*) off the coast of Kent, southeast England', *Journal of Applied Animal Welfare Science* (2010) 13 (1).

21   *Encyclopedia of Marine Mammals*, Second Edition (Oxford, 2008).

22   Mortensen, H. et al.: 'Quantitative relationships in delphinid neocortex', *Frontiers in Neuroanatomy* (2014) 8 (132).

23   Exkursionsbericht des Geographischen Instituts der Uni Bern (1959).
24   www.youtube.com/watch?v=ep2-_ofP19Q.
25   www.diebucht.info.

VIII: Epilogue

1   www.boell.de/sites/default/files/fleischatlas_regional_2016_aufl_3.pdf.
2   www.gesetze-im-internet.de/tierschnutztv/BJNR275800001.html#BJN R275800001BJNG000502308.
3   Manteuffel, G., Schön, P. C.: 'Measuring welfare of pigs by automatic monitoring of stress sounds. Measurement Systems for Animal Data', *Bornimer Agrartechnische Berichte* (2002) (29), pp. 110–118.
4   Kawai, M.: 'Newly-acquired pre-cultural behavior of the natural troop of Japanese monkeys on Koshima Islet', *Primates* (1965) 6, pp. 1–30.
5   Sommer, V., Lowe, A., Dietrich, T.: 'Not eating like a pig: European wild boar wash their food', *Animal Cognition* (2016) 19, p. 245.
6   Marino, L., Colvin, C. M.: 'Thinking pigs: a comparative review of cognition, emotion, and personality in *Sus domesticus*', *International Journal of Comparative Psychology* (2015) 28.
7   www.seeker.com/iq-tests-suggest-pigs-are-smart-as-dogs-chimps-1769934406.html.
8   www.youtube.com/watch?v=mza1EQ6aLdg.
9   Balcombe, J.: 'Animal pleasure and its moral significance', *Applied Animal Behaviour Science* (2009) 118, pp. 208–216.
10   Gintis, H.: 'The evolution of private property', *Journal of Economic Behavior and Organization* (2007) 64, pp. 1–16.
11   Brensing, K.: 'Tödliches Hämmern. Die Gefahren der Windkraft zur See für die Meeresfauna.' In: Etscheit, G.: *Geopferte Landschaften: Wie die Energiewende unsere Umwelt zerstört* (Munich, 2016), pp. 187–205.
12   Raspé, C.: 'Die tierliche Person. Vorschlag einer auf der Analyse der Tier-Mensch-Beziehung in Gesellschaft, Ethik und Recht basierenden Neupositionierung des Tieres im deutschen Rechtssystem' (Berlin, 2013).
13   Brensing, K.: *Persönlichkeitsrechte für Tiere: Die nächste Stufe der moralischen Evolution* (Freiburg, 2013).
14   Sommer, V.: 'Are apes persons? Demanding legal rights for our next of kin', *Folia Primatologica* (2016) 87 (3).
15   Osvath, M.: Spontaneous planning for future stone throwing by a male chimpanzee. *Current Biology* (2009) 19 (5), pp. R190–R191.

Acknowledgements

This book would undoubtedly never have seen the light of day if my wife Katrin, notwithstanding her full-time job as a science journalist, had not picked up so much of the slack for me. I have also lost count of the number of times that our two boys came and stood beside my desk, only for me to shoo them away because I was too busy writing. I solemnly promise you, my nearest and dearest, that I will make it all up to you. Among my colleagues, I am known for my – shall we say – imaginative spelling; however, the best efforts of my mother, a special-school teacher for dyslexia, ensured that my manuscript was brought into line with standard German orthography. Many thanks, Mum ;–)

I would also like to thank Grandma Bine and Grandpa Hans, who unfailingly stepped into the breach when time got too tight for us to look after the kids.

Not least, my thanks are due to my reader Franziska Günther, who for some reason took a shine to my text and stripped it of all its superfluous ballast.

In addition, I would like to express my gratitude to Mr Immanuel Kant, whose 'categorical imperative', which has become such an integral part of our philosophical culture, is encapsulated in the little rhyming dictum with which I prefaced this book. And I call to mind, with very mixed feelings, all the many animals involved in the experiments I have cited. I sincerely hope that, on the basis of our awareness, we humans will learn to treat them more humanely.

Index